**PERGAMON INTERNATIONAL LIBRARY**
of Science, Technology, Engineering and Social Studies
*The 1000-volume original paperback library in aid of education,
industrial training and the enjoyment of leisure*
Publisher: Robert Maxwell, M.C.

# *Introductory Dynamic Oceanography*

**Other titles of interest**

*Books*

FEDOROV:  The Thermohaline Fine-Structure of the Ocean

KRAUS:  Modelling and Prediction of the Upper Layers of the Ocean

McLELLAN:  Elements of Physical Oceanography

PARSONS & TAKAHASHI:  Biological Oceanographic Processes

PICKARD:  Descriptive Physical Oceanography, 2nd edition

TCHERNIA:  Descriptive Regional Oceanography

*Journals*

Deep-Sea Research

Marine Pollution Bulletin

Ocean Engineering

Oceanographic Abstracts and Oceanographic Bibliography

Progress in Oceanography

*Free specimen copies of all journals listed and inspection copies of the books are available on request.*

# Introductory
# Dynamic Oceanography

by

**STEPHEN POND, B.Sc., Ph.D.,**
*Professor*

INSTITUTE OF OCEANOGRAPHY,
UNIVERSITY OF BRITISH COLUMBIA

and

**GEORGE L. PICKARD, M.A., D.Phil., F.R.S.C.,**
*Professor and Director,*

INSTITUTE OF OCEANOGRAPHY,
UNIVERSITY OF BRITISH COLUMBIA

# PERGAMON PRESS

OXFORD · NEW YORK · TORONTO · SYDNEY
PARIS · FRANKFURT

| U.K. | Pergamon Press Ltd., Headington Hill Hall, Oxford OX3 0BW, England |
|---|---|
| U.S.A. | Pergamon Press Inc., Maxwell House, Fairview Park, Elmsford, New York 10523, U.S.A. |
| CANADA | Pergamon of Canada Ltd., 75 The East Mall, Toronto, Ontario, Canada |
| AUSTRALIA | Pergamon Press (Aust.) Pty. Ltd., 19a Boundary Street, Rushcutters Bay, N.S.W. 2011, Australia |
| FRANCE | Pergamon Press SARL, 24 rue des Ecoles, 75240 Paris, Cedex 05, France |
| FEDERAL REPUBLIC OF GERMANY | Pergamon Press GmbH, 6242 Kronberg-Taunus, Pferdstrasse 1, Federal Republic of Germany |

First edition 1978

**British Library Cataloguing in Publication Data**

Pickard, George Lawson
Introductory dynamic oceanography.
1. Ocean currents  2. Ocean waves
I. Title  II. Pond, Stephen
551.4'7       GC201       77-30705
ISBN 0-08-021614-5 hardcover
ISBN 0-08-021615-3 flexicover

*In order to make this volume available as economically and as rapidly as possible the author's typescript has been reproduced in its original form. This method unfortunately has its typographical limitations but it is hoped that they in no way distract the reader.*

*Printed in Great Britain by William Clowes & Sons Limited
London, Beccles and Colchester*

# Contents

# Contents

Contents

x                              Contents

# Preface

The purpose of this text is to present an introduction to Dynamic Physical
Oceanography at a level suitable for senior year undergraduate students in the
sciences and for graduate students entering oceanography.

The aims are to introduce the basic objectives and procedures and to state
some of the present limitations of dynamic oceanography and its relations to
the material of descriptive oceanography. We hope that the presentation will
serve to introduce the field to physicists intending to specialize in physical
oceanography, to help oceanographers in other disciplines to learn enough
about the ocean circulation to discuss with the physical oceanographer the
aspects which he needs to understand for his own work, and to give those in
allied fields an appreciation of what the dynamic oceanographer is trying to
do in contributing to our overall knowledge of the oceans.

The presentation involves the use of mathematics, as the essence of the dynamic
approach is to deduce quantitative information about the movements of the ocean
from mathematical statements of the basic principles of physics. The level is
such that undergraduates who have taken a course in calculus should be able to
follow the essentials of the mathematical arguments, while students in the
physical sciences should have no difficulty at all. Non-physical science
students should not be disheartened by the mathematics because a course with
much of this material has been taken for many years by biological science
students, among others, at The University of British Columbia to complement a
course in descriptive physical oceanography. For students with little calculus
background we emphasize the physical assumptions made in setting up and in
solving the equations, so that the limitations inherent in will be clear, and
then we stress the interpretation of the solutions obtained. The intermediate
mathematical steps are provided for those interested in following them. The
student with limited mathematical background should concentrate on the verbal
physical interpretations and not worry about the details of the equations.
The non-physicist may find parts of Chapter 7 somewhat difficult at first. If
so rereading this chapter after reading Chapters 8, 9 and, perhaps, 10 would
probably be worthwhile.

We have tried to make the text self-contained but we feel that students interested in dynamic oceanography would find it beneficial to acquaint themselves
first with the observational aspects of physical oceanography in order to be
aware of the characteristics of the ocean which the dynamic oceanographer is
endeavouring to understand and explain. A text such as *Descriptive Physical
Oceanography* by Pickard or other introductions to this aspect listed at the
end of the text would provide the desired background.

In assembling the text we have added significantly to the original course
material on which it was based so that it is unlikely that all of the present
material could be covered in a course of twenty-five or so lectures as we
have done in the past. However, we assume that an instructor will select what
is considered appropriate for the class and will leave the remaining material

for later reference or will consider presenting the material in a longer course.

The added material is not only to make the text more complete for the non-physicist but also to try to make the text usable for physical science students who are beginning graduate studies with the intention of pursuing careers as physical oceanographers. The more detailed, extra material has been put toward the ends of chapters when possible. For physical oceanography graduate students the book will need to be supplemented either in lectures or with references to the literature, e.g., further discussion (with more complete mathematical theory) of such topics as turbulence, vorticity, equatorial undercurrents, boundary layers, thermocline and thermohaline circulation theories, etc. The chapters here on waves and tides are for the non-specialist audience as these topics require extensive individual treatment for the physical oceanographer.

If the physicists find that some concepts are introduced in a rather elementary fashion we ask them to bear with us as these are ones which, in our experience, have given trouble to non-physicists. The physicist might even find the more extensive verbal explanations a pleasant relief from the multitudes of equations with limited explanations sometimes encountered.

We have concentrated on the large-scale average circulation to provide a focus for the book. Coastal and estuarine dynamics have not been presented although there is occasional mention of them. The Further Reading list gives some suitable references for these aspects.

In an introductory book we feel that detailed literature references are distracting and so we have not used them extensively. We have used people's names to identify major ideas and occasionally in the specialized sections we have given literature references as starting points for further study. The Suggestions for Further Reading at the end of the book have brief comments to help the reader to judge the level of the works cited.

Finally it must be re-stated that this text is simply an introduction to dynamic oceanography. Much more sophisticated treatments are available for many aspects. Good introductions to the earlier mathematical studies are Stommel's *The Gulf Stream* (1964) and Robinson's compilation of papers (1963). Most of the recent work is still in articles in journals such as those mentioned in the Further Reading list. Even the information on the physical properties of sea water, e.g., density, conductivity, compressibility etc. is being reviewed very carefully and new determinations are being made to satisfy the needs of the increasingly refined analytical treatments being developed.

# List of Main Symbols used in the Text

An underbar added to a symbol indicates that the quantity
is a vector, e.g., $\underline{F}$, $\underline{V}$; the symbol alone indicates the
magnitude of the quantity.

<u>ROMAN LETTERS</u>

| | |
|---|---|
| a | Acceleration |
| A (or $\delta A$) | Area; wave amplitude (Chap. 12) |
| $A_x$, $A_y$, $A_z$, $A_H$ | Kinematic eddy viscosity for x, y, z, horizontal directions (In the latter part of Chap. 9, A is used for brevity for $A_H$ for the transport stream function.) |
| B | Radius of circle (Chap. 8) |
| C | Conductivity (Chap. 2, 5); speed of sound (Chap. 5); speed of wave motion (Chap. 11, 12) |
| $C_d$, $C_s$ | Speed of deep-water, shallow-water waves |
| $C_H$ | Horizontal Coriolis parameter |
| $C_D$ | Aerodynamic drag coefficient |
| d | Relative density (Chap. 2); level of interface of two-layer system (Chap. 9) |
| D | Depth (Chap. 2); geopotential in mixed system units (Chap. 8); thickness of a layer (Chap. 9, 10) |
| $D_E$ | Ekman depth |
| $D_z$ | Particle orbit diameter in wave motion (Chap. 12) |
| exp( ) | The exponential function of ( ) |
| E | Stability (Hesselberg) |
| $E_p$, $E_p^o$ | Potential, standard potential energy of a water column |
| $E_x$, $E_y$, $E_z$, $E_H$ | Ekman numbers |
| f | Coriolis parameter = $2\Omega \cdot \sin\phi$ = planetary vorticity |
| f( ) | Function of ( ) |
| F | Force; with subscripts, a particular force or force component |
| g | Acceleration due to gravity (taken as 9.80 m s$^{-2}$ in this text) |
| $g_f$ | Gravitational attraction of earth on unit mass in an inertial coordinate system |
| G | Gravitational constant |
| h | Water depth; mixed layer depth (Chap. 10) |

| | |
|---|---|
| H | Scale depth (Chap. 4, 7); wave height (Chap. 12) |
| $H_s$ | Significant wave height |
| i | Angle between an isobaric surface and a level (horizontal) surface |
| $\underline{i}, \underline{j}, \underline{k}$ | Unit vectors in the x, y, z directions |
| K | Compressibility (App. 2) |
| $K_x, K_y, K_z, K_H$ | Kinematic eddy diffusivity for the x, y, z, horizontal directions |
| L | Horizontal scale length; wavelength (Chap. 12) |
| $L_b$ | Length of basin (Chap. 13) |
| $L_c$ | Critical length of open-ended resonating water body |
| m, M | Mass |
| $\underline{M}$ | Vector mass transport (per unit width) |
| $M_x, M_y$ | Mass transport (per unit width) in x, y directions |
| M, L, T, K | Within [ ] indicate physical dimensions of mass, length, time, degrees Kelvin |
| n | Integer; normal coordinate (i.e., perpendicular to some surface or line) |
| $n_H$ | Normal coordinate in horizontal plane |
| N | Brunt-Väisälä frequency |
| p | Pressure |
| q | Stands for (any) quantity or variable (App. 1) |
| $Q_T$ | Temperature (heat) source function |
| $Q_x, Q_y$ | Volume transport (per unit width) in x, y directions |
| r | Distance between centres of two masses |
| R | Gas constant (Chap. 2); distance from centre of the earth (Chap. 6, 13) |
| Re, Ri, Ro | Reynolds, Richardson, Rossby numbers |
| s in δs | Element of surface area |
| S | Salinity |
| t | Time |
| T | *In situ* temperature; scale time (Chap. 7); total transport (Chap. 9); period (Chap. 12) |
| $T_f$ | One pendulum day |
| u, v, w | Velocity components in x, y, z directions. Various subscripts are used, e.g., b = barotropic, c = baroclinic, E = Ekman, g = geostrophic |
| U, V, W | Characteristic values for x, y, z velocity components |
| $\underline{V}; \underline{V}_H$ | (Vector) velocity = $\underline{i}\cdot u + \underline{j}\cdot v + \underline{k}\cdot w$; (vector) velocity in a horizontal plane = $\underline{i}\cdot u + \underline{j}\cdot v$ |

$V_b$, $V_c$            Barotropic, baroclinic parts of $V_H$

$V_1$, $V_2$            Horizontal velocity components normal to a vertical section
                   at levels 1 and 2

$V_o$              Speed of Ekman flow at the surface

$V$, $\delta V$            Volume; element of volume

$w_E$              Vertical velocity component at the bottom of the Ekman layer

$W$                Work (Chap. 8); wind speed (Chap. 9); width of western
                   boundary current (Chap. 9, 11)


### GREEK LETTERS

$\alpha$(alpha)          Specific volume

$\beta$(beta)           = $\partial f/\partial y$ = variation of Coriolis parameter with latitude

$\Gamma$(gamma)          Adiabatic temperature gradient

$\delta$(delta)          Specific volume anomaly

$\delta_{T,p}$, $\delta_{S,p}$, $\delta_{S,T,p}$  Components of $\delta$

$\Delta_{S,T}$            Thermosteric anomaly

$\varepsilon_{S,p}$, $\varepsilon_{T,p}$ (epsilon)  Density anomaly terms

$\zeta$(zeta)           Relative vorticity

$\eta$(eta)            Vertical displacement of the surface; as a subscript it
                   indicates a surface value for a quantity

$\theta$(theta)          Potential temperature; angle (App. 1)

$\kappa_S$, $\kappa_T$(kappa)      Kinematic molecular diffusivity for salt, temperature (heat)

$\lambda$(lambda)         Rossby radius of deformation

$\mu$(mu)             Dynamic molecular viscosity

$\nu$(nu)             Kinematic molecular viscosity

$\xi$, $\gamma$(xi,gamma)    Non-dimensional x, y coordinates

$\pi$(pi)             Ratio of circumference to diameter of a circle

$\rho$(rho)            Density

$\sigma_t$(sigma-t)       (Density at atmospheric pressure − 1000) kg m$^{-3}$

$\sigma_\theta$(sigma-$\theta$)       Equivalent of $\sigma_t$ using potential temperature ($\theta$)

$\sigma_w^2$(sigma-w)       Variance of wind speeds

$\tau$(tau)            Frictional stress

$\phi$(phi, lower case)  Geographic latitude

$\Phi$(phi, upper case); $\Delta\Phi_s$; $\Delta\Phi$  Geopotential; standard geopotential 'distance';
                   geopotential anomaly

$\chi$(chi)            Potential energy anomaly

$\psi$(psi)            Stream function

$\Omega$(omega)          Angular speed of rotation of the earth about its axis

MATHEMATICAL SYMBOLS

| | | | |
|---|---|---|---|
| = | equal to | $^-$ (overbar) | Average quantity |
| $\simeq$ | approximately equal to | $_-$ (underbar) | Vector quantity |
| $\sim$ | of the order of | $\Delta x$ (delta-x) | A finite change of x |
| > | greater than | $\delta x$ (delta-x) | A small (finite) change of x |
| >> | much greater than | | |
| < | less than | $\dfrac{du}{dx}$ | Derivative of u with respect to x |
| << | much less than | | |
| $\nabla$ (grad) | gradient operator | $\dfrac{\partial u}{\partial x}$ | Partial derivative of u with respect to x |
| $\nabla .$ (del) | divergence operator | | |
| $\nabla^4$ | biharmonic operator | $\dfrac{du}{dt}$ | Total or individual derivative of u (see (App. 1) |
| \| \| | absolute value of | | |
| $\therefore$ | therefore, $\because$ because | | |

ABBREVIATIONS FOR FREQUENTLY USED UNITS

(SI)    m   metre            kg   kilogram              h   hour
        km  kilometre        s    second               J   Joule

(other):  Sv  =  Sverdrup  =  $10^6$ m$^3$ s$^{-1}$  (volume transport)

# Acknowledgements

We gratefully acknowledge permission from authors and publishers to reproduce figures in Chapters 9, 11 and 12 as indicated in the respective captions.

We are indebted to our colleagues Drs. R.W. Burling, P. Crean, P.H. Leblond and T.R. Osborn for reading parts of the manuscript and offering many helpful suggestions, and to numerous students whose questions have often provoked us into more careful scrutiny of concepts and deductions than we had given them before.

We thank Miss Deina Blackwell for typing numerous drafts of the lecture notes on which this text is based and Mrs. M. Ellis for typing the final manuscript for publication.

Finally we are indebted to the many physicists and oceanographers, past and present, on whose works this book is based.

# CHAPTER 1
# Introduction

*Oceanography* is the study of the ocean making use of the various basic sciences, physics, chemistry, biology and geology, with mathematics being used as an aid to parts of all these studies. Particular attention is paid to the ocean as an environment both for the organisms which inhabit it naturally and in relation to man's activities, and also to its interaction with the atmosphere, the environment in which man lives.

The physicist's contribution is to study the distribution of properties such as temperature, salinity, density, transparency, etc., which distinguish one water mass from another, and to study and understand the motions of the ocean in response to the forces acting.

Some of the problems which have been recognized and studied are:
> Why are the gross mid-latitude surface circulations in the ocean clock-
> wise in the northern hemisphere but anti-clockwise in the southern
> hemisphere?
> Why are these circulations concentrated and swift at the western sides
> (Gulf Stream, Kuroshio, etc.) but broad and slow elsewhere?
> What is the reason for the eastward circulation of the Southern Ocean
> around Antarctica?
> What is the distribution with depth of ocean currents?
> What is (are) the reason(s) for the complicated equatorial flow patterns?
> What are the details of the mechanisms of transfer of momentum and energy
> between air and water?
> What are the characteristics and causes of surface and internal waves?
> What are the relations between submarine earthquakes and tsunamis and
> how do the latter behave in the deep ocean and at the shore?
> What are the characteristics and significances of turbulent motions in
> the oceans?

To some of the questions we have answers, to some we have partial answers, and as our studies progress new problems become apparent.

Physical studies are carried out both by direct observation of the properties and movements and also by applying the basic physical principles of mechanics and thermodynamics to determine the motions. The observational approach is called *descriptive* or *synoptic* oceanography because the physicist tries to reduce his observations to a simple summary or synopsis. The essential feature of the second approach, *dynamic* oceanography, is to use physical laws to endeavour to obtain mathematical relations between the forces acting on the ocean waters and their consequent motions. In either case, the ultimate objective is to learn enough about the structure and motion of the ocean to be able to predict its future state.

In principle, to achieve this objective the dynamic approach is most likely to be successful because it should result in analytic expressions which can be used for prediction into the future, whereas, the synoptic approach simply describes what happened in the past. In practice, it turns out that some characteristics of the oceans do not change much with time, or they repeat themselves with recognized periods, so that a good description of the present state may be applicable for some time into the future. However, some features do change and the amplitudes of cyclic variations may alter, so that a quantitative understanding of the relation between the causative forces and the reaction of the ocean is desirable. Therefore, the preliminary quantitative description of the ocean and its movements prepared by the synoptic ocean-ographer is used by the dynamic oceanographer to suggest what kinds of motion he may expect and what forces may be causing them, so helping him to start his theoretical study. Also, if he meets mathematical difficulties in his analysis, as often happens, the available observations may suggest what mathematical approximations may be made while still keeping the investigation physically realistic.

When the dynamic oceanographer has made a preliminary analysis, it will probably suggest the need for more extensive or sophisticated observations; when these have been made he may refine his analysis. Successions of improved observations and analysis will hopefully lead to better and better under-standing of the physics of the ocean, and improvement in our ability to predict its future state.

Systematic physical observations of the ocean have been made for a century or so, the rate of accumulation of data having increased enormously during the last twenty years. An introduction to the available information is presented in *Descriptive Physical Oceanography* by Pickard and other texts listed in the Suggestions for Further Reading later in this book. The purpose of the pres-ent text is to provide a parallel introduction to dynamic oceanography. The systematic dynamical study of the circulation started at the end of the last century when Scandinavian meteorologists, recognizing the similarity between the dynamics of the atmosphere and of the ocean, turned their attention to the latter. Studies of some phenomena started much earlier with Newton's (circa 1687) and Laplace's (circa 1775) studies of the tides, and Gerstner's (1802) and Stokes' (1874) studies of waves, as examples.

The sequence of topics in this book starts with a description of the proper-ties of sea water relevant to dynamic oceanography, and a summary of the basic physical laws or principles which will be used. The principle of con-servation of mass is used in the form of conservation of volume which, for the incompressibility approximation used, places mild restrictions on the possible motions. An example is given of its application. A definition of static stability is presented and discussed, and the possibility of double diffusive instability is mentioned. Then the forces which may be acting in the sea are classified and some simple examples given. Following this chapter, and occupying a large part of the present exposition is a discussion of the application of Newton's Second Law of Motion to relate the forces acting to the resultant motion of the sea on a rotating earth - the field of geophysical fluid dynamics. Because of analytic (mathematical) difficulties in solving some of the forms of the equations of motion, numerical methods for solving the equations are increasingly being applied; the basic principles, successes and limitations of this technique are summarized. Short accounts of the characteristics of waves and tides are given, and finally some

discussion is offered of what appear to be the presently active and future
areas for research.

Two appendices are included. The first is a brief review of mathematical
techniques used and of simple hydrodynamical principles for the reader with
a limited background in these fields. The second is concerned with units.
In the mathematical solution of equations, when all quantities are represented
by letter symbols, the question of what physical system of units should be
used in measurement does not arise. However, as soon as numerical calculations are to be made, to compare the mathematics with observations made on the
real world, it is necessary to select a system of units. Unfortunately, in
most of the physical oceanographic literature a mixed system of units has
been used. It is basically the CGS system using centimetres, grams, seconds
and calories as the fundamental units. However, although density is expressed
in grams $cm^{-3}$, depths are expressed in metres, and horizontal distances often
in nautical miles, pressure is expressed in decibars, abbreviated as db,
(because the depth in metres and the pressure then have numerically almost the
same value) and a quantity called 'dynamic height' expressed in 'dynamic
metres' is introduced although it is dimensionally work per unit mass. Because
the International System of Units is now coming into general use, and is often
required for publication in many journals, we have elected to depart from the
conventional oceanographic units and introduce the S I units systematically
in this text. Appendix II then contains a glossary of physical oceanographic
terms and conversion factors between the S I units and the old mixed system
units.

For coordinate axes we will use a right-handed system with the positive x-axis
directed horizontally to the east, the positive y-axis horizontally to the
north, and the positive z-axis vertically upward, with the origin normally at
mean sea level. Note that the term 'depth', the distance below the surface,
is taken to be positive as is the usual practice; thus with the origin at the
surface, z is the negative of the depth, i.e., for 'a depth of 100 m' then
$z = -100\,m$.

# CHAPTER 2
# Properties of Sea Water Relevant to
# Physical Oceanography

INTRODUCTION

The physical properties of pure water relevant to fluid dynamics studies are
functions of pressure (p) and temperature (T) while those of sea water are
functions of pressure, temperature and salinity (S).  The salinity of sea
water is a measure of the amount of dissolved salts expressed as the number
of grams of dissolved material in one kilogram of sea water.  The average
value for sea water is about 35 grams per 1000 grams, expressed as S = 35 ‰
(parts per thousand).  Because of the variety of dissolved salts in sea water
and of the physical/chemical problems associated with determining the amount
in a given sample, the exact definition of salinity (given in Appendix II) is
slightly more complicated but as we are not concerned here with the techniques
of determination the above definition will be sufficient.

The effect of the dissolved salts is to alter the physical properties from
those of pure water in degree rather than to develop new properties, e.g.,
small changes in compressibility, thermal expansion, refractivity and larger
changes in the freezing point, density, temperature of maximum density and
electrical conductivity.  Although water is a very common substance it has
extreme values for many physical properties,  e.g., high specific heat so that
ocean currents carry much thermal energy, a high latent heat of fusion so that
in polar regions where there is ice in the water the temperature is maintained
close to the melting point, a high latent heat of evaporation which  is
important in heat transfer from sea to air, and a high molecular heat conductivity.  (This latter property is over-shadowed under most circumstances
by 'eddy' transfer processes due to the turbulent motion of ocean waters.
The 'eddy' or turbulent heat transfer effects are discussed briefly in
Chapter 10.)

DENSITY

From the point of view of dynamic oceanography the most important aspect is
the quantitative manner in which the *density* varies with changes in temperature, salinity and pressure.  Density ($\rho$) decreases as temperature increases,
and increases as salinity and pressure increase.  The variation of density,
as  $\sigma_t$ (= $\rho$ - 1,000) kg m$^{-3}$, is shown in Fig. 2.1 for temperatures from -2 to
30°C and salinities from 30 to 40‰ which cover the ranges of values found in
the open oceans.  About 90 % of the volume of the ocean has values in the much
smaller range from -2° to 10°C and 34 to 35‰ as shown in the figure.  This is
mostly sub-surface water, the remainder of the range of properties in the
figure represents limited volumes of surface waters.  The relation between
density and the parameters temperature and salinity is non-linear, more so

4

Fig. 2.1.   Values of density (as sigma-t) as a function of temperature and
            salinity over ranges appropriate to most of the ocean.   (90% of
            the ocean volume has temperature and salinity values within the
            dashed rectangle.)


in temperature than in salinity, and density is less sensitive to temperature
changes at low temperatures than at high temperatures.   Note that pure water
has a density maximum close to 4°C at atmospheric pressure but as salinity
increases this temperature decreases to about -1.4°C at salinity = 25‰
(where the freezing point has the same value).   A useful rule-of-thumb is that
density increases by approximately 1 part in a thousand (i.e., by 1 kg m$^{-3}$)
for a change of temperature of -5°C, for a change of salinity of +1‰ or for
a change of pressure of +2000 kPa (kiloPascals) (= +200 db equivalent to a
depth change of about +200 m).

Measurement of Density, Temperature and Salinity

For many calculations, the physical oceanographer needs to know the distribution of density both horizontally and vertically in the sea but at present
there is no method available for measuring it accurately *in situ*.  It can be
measured directly in the laboratory but the standard methods are slow.  In
practice, temperature, salinity and pressure are measured and the density is
deduced from tables which have been prepared from laboratory determinations
of this property.

*In situ* temperature (T) is measured either with a specially designed mercury-
in-glass thermometer which records the temperature at the moment when the
sample is taken at depth (reversing thermometer) or with an electrical resistance thermometer.

Because sea water is slightly compressible, a sample brought from depth to
the surface will expand and therefore tend to cool.  The temperature of a
sample brought *adiabatically* to the surface (i.e., without thermal contact
with the surrounding water) will therefore be cooler than *in situ*.  The
temperature which it would have at the surface in these circumstances is
called the *potential* temperature ($\theta$).  This value is used when comparing water
masses at significantly different depths or when considering vertical motions
over considerable depth ranges.

Salinity is determined from measurements of the electrical conductivity and
the temperature of a sample in the laboratory (because at constant pressure
conductivity is a function of temperature and salinity, i.e., conductivity =
f(T,S), a compact way of stating that conductivity is a function of, or varies
with, T and S).  Pressure (p) is determined from the depth of sampling and the
density of the water column.  Alternatively, conductivity, temperature, and
pressure sensors may be mounted together in an underwater unit which is
lowered through the depth range required and the instrument (C,T,D) either
records the data internally or transmits them as electrical signals to
instruments on deck where a continuous record of temperature and conductivity
(or salinity) against depth is obtained.  The relations between salinity,
conductivity and temperature were redetermined in 1966; absolute redetermin-
ations of conductivity and density are still to be made.

Relative Density, Sigma-t and Specific Volume

In their discussions, physical oceanographers sometimes use density ($\rho$), sometimes relative density (d), sometimes specific volume ($\alpha = 1/\rho$) and sometimes
a quantity called 'sigma-t' ($\sigma_t$).

Although oceanographers talk about the 'density' of sea water, strictly speaking they should use the term 'relative density' (formerly 'specific gravity')
because the laboratory determinations on sea water of the effects of temperature and salinity have consisted of comparisons with pure water and the
direct determinations of $\rho = \rho(T,S,p)$ are less precise.  The relative density
can be given with an accuracy of about 3 in $10^6$ but the density is only
accurate to about 10 in $10^6$.  Fortunately it is the density differences which
are important in most cases and for these the greater uncertainty in the
absolute values is not significant.  We will follow the common practice and
talk about $\rho$ as density with the physical dimensions $[ML^{-3}]$.

It should be noted that in oceanographic practice, when specifying the conditions for a water sample the quantity p refers to the hydrostatic pressure, i.e., the pressure due only to the column of water above a point in the sea, so that p = 0 implies that the sample is at atmospheric pressure.

*Sigma-t*, defined as $\sigma_t = (\rho_{S,T,0} - 1,000)$, was introduced simply for brevity. The density of sea water at atmospheric pressure varies from about 1,000 kg m$^{-3}$ (1.000 g cm$^{-3}$) for almost fresh water to about 1,028 kg m$^{-3}$ for the densest ocean surface water. As the variation is entirely in the last two figures (four if density is expressed to the second decimal place) it is convenient for descriptive purposes simply to use these two (or four) figures, e.g., for sea water of T = 10.00°C, S = 35.00°/oo and p = 0, then $\rho$ = 1,026.97 kg m$^{-3}$ and $\sigma_t$ = 26.97 kg m$^{-3}$. Although $\sigma_t$ has units of kg m$^{-3}$ it is usual to omit them when quoting values.

It should also be noted that $\sigma_t$ is the special case of the more general *in situ* quantity $\sigma_{S,T,p} = (\rho_{S,T,p} - 1,000)$ which includes the effect of pressure but is less often used than $\sigma_t$. Sigma-t is more often used because it allows a much better estimate of what the density difference between two water types would be when they are at the same level and hence it is a better indicator of static stability (which will be discussed in Chapter 5). When one is considering the motion of a water parcel over a considerable depth range, it may be desirable to eliminate the effect of adiabatic heating or cooling by using the potential temperature $\theta$ rather than the *in situ* temperature T and calculating the potential density $\rho_{S,\theta,0}$ or $\sigma_\theta = (\rho_{S,\theta,0} - 1,000)$.

*Specific volume* ($\alpha$) is the reciprocal of density ($\alpha = 1/\rho$) and has the units m$^3$ kg$^{-1}$. Two other density related quantities are used in oceanography. These are specific volume anomaly ($\delta$) and thermosteric anomaly ($\Delta_{S,T}$) which will be defined later in this chapter. The relations between the units for density and specific volume, *etc.* used here and those used in previous texts are discussed later in the chapter and also in Appendix II for reference.

For orientation and for comparison, some values for density for fresh water and for sea water will be given. For the open ocean, temperatures range from -2°C to +30°C, salinity from 30 to 38 °/oo and hydrostatic pressure from 0 to about 10$^5$ kPa (= 10$^4$ db, corresponding to the maximum ocean depth of about 10,000 m). Near rivers or melting ice, the salinity may fall to 0 °/oo in the surface layer, while values of 40 °/oo or more occur in the Red Sea. Values for density are given in Table 2.1.

The density of fresh water at 0 kPa hydrostatic pressure (i.e., at standard atmospheric pressure) has its maximum value of 999.97 kg m$^{-3}$ at 3.98°C. Values in brackets in Table 2.1 correspond to conditions which do not occur in lakes or oceans.

## TABLES FOR DENSITY AND SPECIFIC VOLUME AS FUNCTIONS OF TEMPERATURE, SALINITY AND PRESSURE

For a one-component fluid, the specific volume $\alpha = \alpha(T,p)$, i.e., it is a function of temperature and pressure only. For a perfect gas, the relation (the 'equation of state') has the simple form $\alpha = R.T/p$ where R is the Gas Constant. For fresh water, the relationship is more complicated, while for sea water, which is a multi-component fluid, the dissolved salts add further complication and $\alpha = \alpha(S,T,p)$. (The situation is similar in the atmosphere

TABLE 2.1  Values of Density *In Situ* for
Fresh and Sea Water (kgm$^{-3}$)

| Hydrostatic Pressure | Approx. Depth | | Fresh Water S = 0‰ | | Average Sea Water S = 35‰ | | Red Sea (winter) S = 40‰ |
|---|---|---|---|---|---|---|---|
| | | | Temperature | | Temperature | | Temperature |
| $10^4$kPa $10^3$db | m | | 0° | 30°C | 0° | 30°C | 18°C |
| 0    0 | 0 | | 999.8 | 995.9 | 1028.1 | 1021.8 | 1029.1 |
| 1    1 | 1,000 | | 1004.8 | (1000.0) | 1032.8 | (1026.0) | 1033.5 |
| 4    4 | 4,000 | | (1019.3) | (1012.8) | 1046.3 | (1038.1) | 1045.9 |
| 10   10 | 10,000 | | (1045.4) | (1036.1) | 1070.9 | (1060.5) | (1068.5) |

where water vapour plays an analogous role to salinity in sea water and changes the relatively simple equation of state for dry air to a more compli- cated form for moist air.)  The relationship $\alpha = \alpha(S,T,p)$ can be presented as a complicated polynomial in temperature, salinity and pressure but it is usually expressed in the form of tables.

The early oceanographers who measured the variation of density with ·tempera- ture, salinity and pressure found, by a process of trial and error, that the most convenient way to express the results was in terms of the *specific volume* ($\alpha = 1/\rho$) as follows:

$$\alpha_{S,T,p} = \alpha_{35,0,p} + \delta_S + \delta_T + \delta_{S,T} + \delta_{S,p} + \delta_{T,p} + \delta_{S,T,p}. \qquad (2.1)$$

or as

$$(\alpha_{S,T,p} - \alpha_{35,0,p}) = \delta = \Delta_{S,T} + \delta_{S,p} + \delta_{T,p} + \delta_{S,T,p} .$$

In these expressions, $\alpha_{S,T,p}$ is the specific volume of a sample of water of salinity S, temperature T and pressure p (the hydrostatic pressure at the depth of the sample in the sea).  $\alpha_{35,0,p}$ is the specific volume of sea water of S = 35‰, T = 0°C and pressure p at the depth of the sample.  This term expresses most of the effect of pressure on specific volume.  The term $\delta$, the *specific volume anomaly*, represents the sum of the six anomaly terms in equation 2.1.  The quantity $\Delta_{S,T} = \delta_S + \delta_T + \delta_{S,T}$ accounts for most of the effect of salinity and temperature, disregarding pressure, and is called the *thermosteric anomaly*.  The terms $\delta_{S,p}$ and $\delta_{T,p}$ account respectively for most of the combined effect of salinity and pressure and of temperature and pres- sure.  The last anomaly term, $\delta_{S,T,p}$, is so small that it is always neglected with the present accuracy of determination of the parameters S, T and p.  In water of depth less than about 1,000 m the thermosteric anomaly, $\Delta_{S,T}$, is the major component of $\delta$ and the pressure terms $\delta_{S,p}$ and $\delta_{T,p}$ may often be neglected.  $\Delta_{S,T}$ has, in recent years, to a large extent replaced $\sigma_t$ as a parameter for describing density characteristics in the upper layer of the ocean because it can be used more directly than $\sigma_t$ in first order dynamic calculations.

From    $\alpha_{S,T,0} = 1/\rho_{S,T,0} = 1/(1000 + \sigma_t)$   and   $\alpha_{S,T,0} = \alpha_{35,0,0} + \Delta_{S,T}$

and noting that $\alpha_{35,0,0} = 0.97264 \cdot 10^{-3} \, m^3 kg^{-1}$ it is easy to show that

$$\Delta_{S,T} = \left( \frac{1000}{1000 + \sigma_t} - 0.97264 \right) \cdot 10^{-3} \, m^3 kg^{-1}.$$ A few values are as follows:

$$\sigma_t = 23.00 \quad 24.00 \quad 25.00 \quad 26.00 \quad 27.00 \quad 28.00 \; kg \; m^{-3}$$

$$\Delta_{S,T} = 487.7 \quad 392.2 \quad 297.0 \quad 201.9 \quad 107.0 \quad 12.3 \times 10^{-8} \, m^3 \, kg^{-1}$$

## Sources of Data

Sources of tables of data are as follows (see references at end of text):

| | | |
|---|---|---|
| $\alpha_{35,0,p}$ | – Sverdrup $et$ $al.$, Table I | for p = 0 to 9,900 db (99,000 kPa) |
| | Neumann & Pierson, Table IV | ditto |
| | N.O.O. 614, Table IV | ditto |
| $\Delta_{S,T}$ | – Sverdrup $et$ $al.$, Table III | for $\sigma_t$ = 23 to 28 |
| | Neumann & Pierson, Table I | ditto |
| | N.O.O. 614, Table V | for T = -1.9 to 29.9°C |
| | | S = 21.0 to 37.9 ‰ |
| $\delta_{T,p}, \delta_{S,p}$ | – Sverdrup $et$ $al.$, Tables IV & V | |
| | Neumann & Pierson, Tables II & III | |
| | N.O.O. 614, Tables VI and VII | |

|  | | T | S |
|---|---|---|---|
| $\sigma_t$ | – Knudsen's Tables | for -2 to 33°C | 0 to 40 ‰ |
| | N.O.O. 614 | -2 to 30 | 30 to 38 |
| | N.O.O. 615 | -2 to 30 | 0 to 40 |
| | Fleming (S & T values for unit values of $\sigma_t$) | -2 to 30 | 22 to 41 |
| $\rho_{S,T,p}$ | – reciprocal of $\alpha_{S,T,p}$. | | |

It must be noted that the above sets of tables use the old mixed units system. Using primed symbols (e.g., $\alpha'$) to represent mixed units numerical values as in the tables, and unprimed symbols (e.g., $\alpha$) for SI numerical values, then:

$$\alpha = \alpha' \times 10^{-3} \qquad\qquad \rho = \rho' \times 10^3$$
$$\Delta_{S,T} = \Delta'_{S,T} \times 10^{-3} \qquad\qquad \sigma_t = \sigma'_t$$
$$\delta_{T,p} = \delta'_{T,p} \times 10^{-3} \qquad\qquad p(kPa) = p'(db) \times 10$$
$$\delta_{S,p} = \delta'_{S,p} \times 10^{-3} \qquad\qquad T \; and \; S \; are \; the \; same.$$

To indicate their order of magnitude, Table 2.2 gives a selection of values for the specific volume anomaly terms. Blanks in the table indicate that the combinations of parameters do not occur in the sea.

TABLE 2.2   A Selection of Values for Specific Volume
Anomaly Terms in Units of $10^{-8}$ $m^3 kg^{-1}$.

| $\Delta_{S,T}$ : | Temperature | Salinity: | 30 | 32 | 34 | 35 | 36‰ |
|---|---|---|---|---|---|---|---|
| | -2°C | $\Delta_{S,T}$ = | 378 | 224 | 70 | -7 | -84 |
| | 0 | | 382 | 229 | 76 | 0 | -76 |
| | 10 | | 480 | 331 | 183 | 109 | 36 |
| | 20 | | 681 | 535 | 390 | 318 | 245 |

| $\delta_{T,p}$ : | Temperature | Pressure:<br>(Depth ≈ | 0<br>0 | 1<br>1,000 | 2<br>2,000 | 5<br>5,000 | $10 \times 10^4$ kPa<br>10,000 m) |
|---|---|---|---|---|---|---|---|
| | -2°C | $\delta_{T,p}$ = | 0 | -6 | -11 | | |
| | 0 | | 0 | 0 | 0 | 0 | 0 |
| | 2 | | 0 | 6 | 10 | 22 | 38 |
| | 10 | | 0 | 21 | 41 | | |

| $\delta_{S,p}$ : | Salinity | Pressure:<br>(Depth ≈ | 0<br>0 | 1<br>1,000 | 2<br>2,000 | 5<br>5,000 | $10 \times 10^4$ kPa<br>10,000 m) |
|---|---|---|---|---|---|---|---|
| | 30‰ | $\delta_{S,p}$ = | 0 | -8 | | | |
| | 34 | | 0 | -2 | -3 | -7 | |
| | 34.8 | | 0 | 0 | -1 | -1 | -2 |
| | 35 | | 0 | 0 | 0 | 0 | 0 |
| | 36 | | 0 | 2 | 3 | | |

Table 2.3 shows in the first column the current accuracy of routine measurement of the three parameters, salinity (from conductivity), temperature and
pressure, and in the second and third columns respectively the accuracy of
density and of specific volume corresponding to the variation of each parameter individually.    For comparison the values for density, etc. for water
of temperature 10°C and salinity 35.00‰ at zero hydrostatic pressure are
given at the bottom of the table.

TABLE 2.3   Accuracy of Measurement of Temperature, Salinity and Pressure
and Related Accuracies of Density and Specific Volume

| Accuracy of Measurement | Related Accuracies of | |
|---|---|---|
| | Density in $kg\ m^{-3}$ | Specific Volume in $m^3\ kg^{-1}$ |
| $\Delta S = \pm\ 0.003\%$ | $\Delta\rho = \Delta\sigma_t = \pm\ 0.002$ | $\Delta\alpha = \Delta(\Delta_{S,T}) = \mp\ 0.2 \times 10^{-8}$ |
| $\Delta T = \pm\ 0.02\ C°$ | $\Delta\rho = \Delta\sigma_t = \mp\ 0.003(1)$ | $\Delta\alpha = \Delta(\Delta_{S,T}) = \pm\ 0.3 \times 10^{-8}$ |
| $\Delta p = \pm\ 50\ kPa$ | $\Delta\rho \quad\quad \pm\ 0.024(2)$ | $\Delta\alpha = \quad\quad \mp\ 2.2 \times 10^{-8}$ |

$(\equiv \Delta z = \pm\ 5\ m,\ \text{for upper } 1,000\ m)$

For water of:   T = 10.00°C       then       $\rho$   = 1,026.97 $kg\ m^{-3}$
S = 35.00‰        $\sigma_t$  =    26.97 $kg\ m^{-3}$
p = 0        $\alpha$   =    0.97374 × $10^{-3}\ m^3\ kg^{-1}$
$\Delta_{S,T}$ =   109.7     × $10^{-8}\ m^3 kg^{-1}$.

Notes:  1.  This value is for T = 10.00°C; values for S = 35.00‰ range from
$\Delta\sigma_t = \mp\ 0.002$ at T = 2°C to $\mp$ 0.006 at T = 25°C.

2.  The uncertainty due to pressure differences may be greater in deep
water because of greater uncertainty in pressure measurement.

# CHAPTER 3

# The Basic Physical Laws used in
# Oceanography and Classifications of
# Forces and Motions in the Sea

## BASIC LAWS

The following basic laws of physics are taken as axiomatic in developing the study of the dynamics of the ocean:

(1)  Conservation of mass,
(2)  Conservation of energy,
(3)  Newton's First Law of Motion that if there is no resultant force acting on a body, there will be no change of motion of the body,
(4)  Newton's Second Law of Motion that the rate of change of motion of a body is directly proportional to the resultant force upon it and is in the direction of that force,
(5)  Newton's Third Law of Motion that for any force acting on a body there is an equal and opposite force acting on some other body,
(6)  Conservation of Angular Momentum,
(7)  Newton's Law of Gravitation.

Strictly speaking, the first two are related but in dynamic oceanography we are not concerned with the Mass $\updownarrow$ Energy conversion and therefore it is convenient to keep them separate. Conservation of mass is fundamental but in oceanography it is usually used in the form called the 'equation of continuity' which actually expresses conservation of either mass/unit volume (density) or of volume.

The two types of energy whose conservation is important in oceanography are heat and mechanical. Conservation of heat or the heat budget is most important when discussing the distribution of temperature as a property of the ocean waters in descriptive oceanography and an account of this subject may be found in *Descriptive Physical Oceanography* by Pickard (see the Further Reading list at the end of the text). Conservation of mechanical energy will be considered when treating waves, while the conversion of mechanical to heat energy will be taken for granted as a loss process for the former but will not be discussed in any detail in this text but left for more advanced treatments. (This source of heat is negligible in the heat budget.)

Dynamic oceanography is concerned with the forces acting on the ocean waters and with the motions which ensue. In some cases the motions occur under a system of forces which are in balance so that no resultant force acts - this is the case covered by Newton's First Law of Motion. In other cases there is a resultant force and acceleration occurs, the relations between them being determined by Newton's Second Law. In this text, except in the case of waves and tides, we will be concerned almost entirely with unaccelerated motion,

i.e. with applications of the First Law, in interpreting the dynamic behavior
of the oceans.

We will not use angular momentum directly in this treatment but rather a
related quantity called vorticity. It should be noted that both linear and
angular momentum may not be conserved when we measure them relative to the
earth because the latter is itself rotating in space and the effects of
rotation must be taken into account in the development of the equations of
motion of the ocean waters.

The Law of Gravitation, as such, is principally applied in discussing the
dynamics of the astronomical tides of the ocean, although it is also important
in determining the hydrostatic pressure distribution and in causing motion
when density changes occur.

## CLASSIFICATION OF FORCES AND MOTION

The important forces can be divided into two classes, *primary* which cause
motion, and *secondary* which result from motion. The primary forces are
(1) *gravitation,* both terrestrial, including pressure forces, and due to the
sun and the moon, (2) *wind stress* which may be tangential (friction) and
normal (pressure) to the sea surface, (3) *atmospheric pressure,* and (4) *seismic*
(from sea bottom movements). Gravitation is a body force in that it acts on
the total mass of the water, while the others are boundary forces, acting
first at a water surface, although their effects may penetrate beyond the
surface into the body of the water.

The secondary forces which come into being when water starts to move are
(5) *Coriolis force,* an apparent force on a moving body when its motion is
observed relative to the rotating earth and (6) *friction* acting at the boundary of the fluid and tending to oppose its motion or acting within the fluid
and tending to make the motion more uniform. Friction also tends to dissipate
the mechanical energy of the motion, converting the kinetic energy of the
fluid into heat energy. Again, Coriolis force is a body force whereas boundary friction is initially a surface acting force whose effects penetrate in
diminishing degree into the body of the fluid.

A common classification of motion is as (1) *thermohaline* motion which results
when the density of water changes in a limited region so that the differential
action of gravity causes relative motion. The density changes will be due to
changes of temperature and/or salinity, hence 'thermohaline' motions.
(Thermosaline might appear to be the more logical term but 'thermohaline' is
the more common one and is etymologically correct. It dates back to the
earlier days of oceanography when the salt content was determined by a silver
nitrate titration procedure which determined the total *halogen* content of the
sea water sample, and the salinity was then determined from this quantity.)
Evaporation, cooling and freezing (which raises the salinity of the unfrozen
water) all increase the density of sea water and may cause it to sink vertically, with subsequent horizontal movement at its own density level or along
the bottom, (2) *wind-driven* motions such as the major ocean circulations in
the upper layers, surface waves and upwelling, (3) *tidal* currents which are
essentially horizontal, and internal waves of tidal period, (4) *tsunami* or
*seismic* sea waves resulting from movements of the sea bottom during undersea
earthquakes, (5) *turbulent* motions resulting from velocity shear (change of
velocity with respect to one or more spatial coordinates), usually at the

water boundaries, and (6) various motions such as *internal waves, inertial waves, Rossby or planetary waves,* etc.

It would also be possible to classify motions by their scale (from the small- est turbulent eddies of millimetre scale to the dimensions of the major ocean circulations), by speed, or by the method of determination, but the classifi- cation above in terms chiefly of the causal forces is the most used.

A few comments may be made on the various forces acting on the sea, before we proceed to discuss the relations between forces and motion.  Terrestrial gravity gives rise to the property possessed by mass near the earth which we call weight, and there are two consequences.  In the first place, weight gives rise to the phenomenon of hydrostatic pressure in a fluid.  In the second place, because a fluid has weight there will be a component down slope if a fluid surface is not horizontal and as a fluid cannot withstand shear strain it will tend to flow down slope (see Appendix I).

The astronomical gravitational forces due to the sun and moon fluctuate periodically as the earth rotates and as these two bodies circulate around the earth, and give rise to the periodic motions which we call tidal currents and tides.  (The gravitational forces due to the other bodies in the solar system are negligible compared to those of the earth's moon and of the sun.)

The wind, in blowing relative to the water surface, transfers momentum and energy to the upper layer of water giving rise both to the fluctuating motions of waves and to the steadier ocean currents.  The generation and development of waves involves fluctuations of the normal stress (pressure) of short period of the order of seconds, and perhaps fluctuations of the tangen- tial stress on the water surface, but the ocean currents express an integrated (averaged) effect of the stress.

Differences in atmospheric pressure can cause differences in water levels (called the inverted barometer effect) and hence currents.  For instance at the centre of a cyclonic storm the air pressure is low and the water level tends to rise; there is an inflow of water to raise the level.  Then as the storm moves, these patterns tend to follow and may cause inundations of low- lying coastal areas.

Sudden movements of the ocean bottom, such as occur in a dip-slip earthquake when there is a component of motion perpendicular to the bottom, cause corresponding movements of the sea surface above.  If, for instance, a hump is formed on the sea surface the water immediately begins to flow outward from the hump and a train of waves develops and moves outward from the earth- quake location.  These are tsunamis or seismic sea waves.

The 'Coriolis force' is an apparent force which acts perpendicular to the velocity vector of a moving body on the surface of the earth.  It is the name given to a term which appears when the equation of motion is transformed from a fixed frame of reference (relative to the 'fixed' stars) to a frame of ref- erence fixed in the earth which is itself rotating about its axis.  This force will be discussed in Chapter 6.

The secondary frictional forces are not physically of a different nature from the primary ones but are introduced as secondary forces because they do not arise until motion has been generated, and they then usually tend to oppose the motion rather than maintain it.  Frictional effects arise because of the molecular nature of the fluid and may be much enhanced if the flow is turbulent.

# CHAPTER 4

# The Equation of Continuity of Volume

## THE CONCEPT OF CONTINUITY OF VOLUME

It was stated in the previous chapter that the law of conservation of mass is used in oceanography in the form of an equation of continuity (of volume or density). Before deriving this equation we will consider the physical significance of volume continuity.

For a stationary fluid the idea of continuity of volume is trivial, so let us consider a moving fluid which is assumed to be incompressible (i.e., the volume is not affected by pressure, see Appendix I) and uniform in kind. (Sea water is not exactly incompressible but this assumption is a good approximation for many applications because the volume changes are small. For instance, a change of pressure corresponding to a change of depth of 1,000 m would change the volume of a sample of average sea water by less than 0.5%. There are few situations in ocean currents where depth (and therefore pressure) changes as large as this occur along the flow path.)

Suppose that we had a hand basin with two taps, a waste pipe at the bottom and an overflow pipe near the top. With the taps full on (and a good flow of water from them) it is possible that the water level would rise in the basin. We may reach a final state when the basin fills completely and water starts to drip over the edge - then we will have continuity of volume of water in the basin, which could be expressed as:

$$
\begin{matrix} \text{Hot} \\ \text{inflow} \end{matrix} + \begin{matrix} \text{Cold} \\ \text{inflow} \end{matrix} = \begin{matrix} \text{Waste} \\ \text{outflow} \end{matrix} + \begin{matrix} \text{Overflow} \\ \text{pipe outflow} \end{matrix} + \begin{matrix} \text{Drip out} \\ \text{over edge} \end{matrix}
$$

or as

$$
+ \text{ Hot } + \text{ Cold } - \text{ Waste } - \text{ Overflow } - \text{ Drip } = 0 \ .
$$

This is rather a domestic example. Consider now the following. An oceanographer is studying a long, narrow coastal inlet which has a river at the inland end. He notes that the upper layer is flowing to seaward and that the thickness of this layer is constant (within the observational errors of 10-20%) from river to sea. He also notes that the seaward flow is slow at the river end of the inlet but becomes substantially faster toward the seaward end. He wonders why, because this behaviour implies that for a section of the inlet between AB and A'B' (Fig. 4.1a), where the water speed increases from $u_3$ to $u_4$, there must be more volume of water flowing out across A'B'C'D' than in across ABCD. If we consider the horizontal flow only, there appears to be a lack of continuity of volume. However, the inlet is not emptying and to produce a balance there must be an upward flow (w) from the lower to the upper

(a)  PLAN VIEW

RIVER

$u_1$  $u_2$  $u_3$  $u_4$  SEA

(b)  PERSPECTIVE VIEW

$u_3$  $u_4$  SEA

w

Fig. 4.1.  Continuity of volume for an inlet: (a) plan view of upper layer,
outflow increases seaward, (b) perspective view - outflow in
upper layer, inflow in lower layer and upward flow from lower to
upper layers.

layer across CDD'C' (Fig. 4.1b) so that we have:

$$u_3 \times \text{area ABCD} + w \times \text{area CDD'C'} = u_4 \times \text{area A'B'C'D'}$$

or

$$u_3 \times \text{area ABCD} + w \times \text{area CDD'C'} - u_4 \times \text{area A'B'C'D'} = 0 \ ,$$

expressing continuity of volume for the upper layer.  The water which leaves
the lower layer must be replaced.  Because there can be no flow through the
bottom of the inlet an inward horizontal flow from the sea must develop as
shown.  This is an example of estuarine flow.  Note that this is an idealization of the average behaviour in an inlet.  Real inlets are often more complicated, e.g., the return flow may be laterally adjacent to the outflow,
there may be more than one region of outflow and/or inflow in a cross-section,
and wind-driven and tidal motions may be large enough to make observations of
the average flow extremely difficult.  However, the principle remains valid
that there must be replacement of the deeper water which the river flow picks
up and carries out of the inlet.

THE DERIVATION OF THE EQUATION OF CONTINUITY OF VOLUME

We will now consider conservation of mass in order to derive a general
equation for applying continuity of volume.  In Fig. 4.2 is represented a
rectangular volume fixed in space with sides of lengths $\delta x$, $\delta y$ and $\delta z$ in a

Fig. 4.2.    Continuity of volume - components of flow in x-direction.

moving fluid.  Consider first the flow parallel to the x axis.  At the left
face the velocity is u and the fluid density $\rho$, while at the right face the
velocity is u + $\delta$u and the density $\rho$ + $\delta\rho$.   These last two expressions may
be written approximately as u + $(\partial u/\partial x) \cdot \delta x$ and $\rho$ + $(\partial\rho/\partial x) \cdot \delta x$ where terms
in $(\delta x)^2$ etc. have been neglected because they will vanish when we take the
limit as $\delta x \to 0$.   (A reader unfamiliar with the notation being used should
read Appendix I up to the section on hydrostatic pressure before proceeding
further in the text.)    Then, in the x-direction:

the mass flow into the volume  =  $\rho \cdot u \cdot \delta y \cdot \delta z$ (mass/unit time)

and   the mass flow out of the volume  =  $(\rho + \dfrac{\partial\rho}{\partial x} \cdot \delta x)(u + \dfrac{\partial u}{\partial x} \cdot \delta x) \cdot \delta y \cdot \delta z$

so that the net flow out of the volume in the x direction is the difference

$$[u \cdot \frac{\partial\rho}{\partial x} + \rho \cdot \frac{\partial u}{\partial x} + \frac{\partial\rho}{\partial x} \cdot \frac{\partial u}{\partial x} \cdot \delta x] \cdot \delta x \cdot \delta y \cdot \delta z = [\frac{\partial(\rho u)}{\partial x} + 0(\delta x)] \cdot \delta x \cdot \delta y \cdot \delta z .$$

Here, $0(\delta x)$ has been written for the last term in the brackets of the previous
equation and indicates that the term is of the order (size) $\delta x$ times some
finite number.  By taking $\delta x$ sufficiently small (mathematically we take the
limit as $\delta x \to 0$) this term must become negligible compared with the other term
in the square bracket provided that the multiplier $(\partial\rho/\partial x) \cdot (\partial u/\partial x)$ is finite
as we expect it to be for a physical system.*

Then, taking into account the mass flow in all three component directions and
neglecting the terms which vanish in the limit as $\delta x$, $\delta y$, $\delta z \to 0$ the total

flow out  =  $[ \dfrac{\partial(\rho u)}{\partial x} + \dfrac{\partial(\rho v)}{\partial y} + \dfrac{\partial(\rho w)}{\partial z} ] \cdot \delta x \cdot \delta y \cdot \delta z$

---

*  The reader may be concerned about treating a medium actually made up of
molecules as continuous and about whether taking the limit as $\delta x \to 0$ is realistic.  In fact, there are no practical problems in this limiting process.
For further discussion, which is beyond the scope of this book, the reader
may consult a text on continuum or fluid mechanics, such as that by Batchelor
(1967) in the Further Reading list at the end of this book.

where v and w are the velocity components in the y and z directions respectively.

The mass remaining in the small volume $\delta x \cdot \delta y \cdot \delta z$ changes by $(\partial \rho / \partial t) \cdot \delta x \cdot \delta y \cdot \delta z$ per unit time. If mass is to be conserved, the sum of the effects must be zero, i.e., the change of mass per unit volume is

$$\frac{\partial \rho}{\partial t} + \frac{\partial (\rho u)}{\partial x} + \frac{\partial (\rho v)}{\partial y} + \frac{\partial (\rho w)}{\partial z} = 0 \ . \tag{4.1}$$

Now the rate of change of density with the moving fluid (the individual derivative, see Appendix I) is

$$\frac{d\rho}{dt} = \frac{\partial \rho}{\partial t} + u \cdot \frac{\partial \rho}{\partial x} + v \cdot \frac{\partial \rho}{\partial y} + w \cdot \frac{\partial \rho}{\partial z} \ . \tag{4.2}$$

Combining equations 4.1 and 4.2 we have:

$$\frac{1}{\rho} \cdot \frac{d\rho}{dt} + \left[ \frac{\partial u}{\partial x} + \frac{\partial v}{\partial y} + \frac{\partial w}{\partial z} \right] = 0 \ . \tag{4.3}$$

This is called the equation of continuity (of volume). The first term is the fractional rate of change (change/unit time) of density for a small piece or parcel of the moving fluid (a 'fluid element'); the second term is the fractional rate of change of volume for the element as we shall show in a moment. The equation expresses conservation (continuity) of volume, i.e., the relation between volume and density changes. Notice that we do *not* assume that the fluid has the same density everywhere (a homogeneous fluid), which is important in application to the ocean whose water is not homogeneous. The effects of pressure and heat exchange are included in equations 4.1 and 4.3 since they do not affect the mass of a fluid element appreciably. Salt exchange effects are not included; if we assume that an element exchanges an equal mass of salt and water equation 4.3 gives the correct relation between density and volume changes. However, in the more likely case that about the same number of molecules of water are replaced by salt ions the mass increases and the volume will not decrease as much as 4.3 predicts. In the ocean the effect is small enough to ignore when considering mass or volume conservation. In fact all of these effects are quite small; here they may be ignored and the volume of a fluid element assumed to be constant, i.e., the fluid may be treated as effectively incompressible.

If a fluid is incompressible, as may be taken to be the case for sea water in most circumstances, then $(1/\rho) \cdot (d\rho/dt) = 0$, as shown in Appendix I, and the equation of continuity becomes:

$$\frac{\partial u}{\partial x} + \frac{\partial u}{\partial y} + \frac{\partial w}{\partial z} = 0 \ . \tag{4.4}$$

We note that in deriving the equation of continuity, instead of considering the rate of change of mass in a volume $\delta x \cdot \delta y \cdot \delta z$ fixed in space we could consider the rate of change of volume of a fluid element. If we consider the volume $\delta V = \delta x \cdot \delta y \cdot \delta z$ of Fig. 4.2 to be moving with the fluid, then in time $\delta t$ (eventually we take the limit as $\delta t \to 0$ so we neglect terms proportional to $(\delta t)^2$ and higher powers) side 1 moves $u \cdot \delta t$ while side 2 moves $[u + (\partial u / \partial x) \cdot \delta x] \cdot \delta t$. The change in volume is $(\partial u / \partial x) \cdot \delta x \cdot \delta y \cdot \delta z \cdot \delta t$.

Fig. 4.3.  Sign the of velocity gradient.

Consideration of movements in the y and z directions as well gives the rate of change of the volume $\delta V = \delta x \cdot \delta y \cdot \delta z$ as $d(\delta V)/dt = \delta V \cdot (\partial u/\partial x + \partial v/\partial y + \partial w/\partial z)$. This rate of change of volume must be balanced by a corresponding rate of change of density, because the mass, $\delta m$, of the fluid element must be constant (assuming salt exchange effects are negligible or balance) although its shape, density and volume may change.  The signs are opposite as an increase in volume decreases $\rho$ and *vice versa*.  Now $\rho = \delta m/\delta V$ and with the mass constant

$$\frac{d\rho}{dt} = \frac{d}{dt}\left(\frac{\delta m}{\delta V}\right) = -\frac{\rho}{\delta V}\cdot\frac{d(\delta V)}{dt} = -\rho\left(\frac{\partial u}{\partial x} + \frac{\partial v}{\partial y} + \frac{\partial w}{\partial z}\right) \quad \text{as before.}$$

It may be noted that terms like $\partial u/\partial x$ or $\delta u/\delta x$ may be positive or negative. In Fig. 4.3, case (a), fluid is flowing from A where its velocity component is u to B where it is (u + $\delta$u).  Then along the flow direction, $\delta u$ is +, $\delta x$ is + and therefore $\delta u/\delta x$ is +.   In case (b), the velocity decreases from C to D and $\delta u$ is -, $\delta x$ is + and therefore $\delta u/\delta x$ is -.   In case (c), the flow is to the left from F to E, $\delta u$ is -, $\delta x$ is -, and therefore $\delta u/\delta x$ is +.

### AN APPLICATION OF THE EQUATION OF CONTINUITY

As an example of the application of the equation of continuity, we consider the determination of vertical velocities in the open ocean.  These are difficult to measure directly because they are very small in magnitude, but some information about them may be deduced with the aid of the equation of continuity from a knowledge of horizontal velocities which are larger and more easily measured.

We have     $\dfrac{\partial u}{\partial x} + \dfrac{\partial v}{\partial y} + \dfrac{\partial w}{\partial z} = 0$

i.e.,        $\dfrac{\partial w}{\partial z} = -\left[\dfrac{\partial u}{\partial x} + \dfrac{\partial v}{\partial y}\right]$

Fig. 4.4.  Example of horizontal flows for calculation of vertical flow from continuity of volume. (speeds in $ms^{-1}$)

and from data about u and v (horizontal components) we may learn something about w (vertical), near a point in the sea such as E in Fig. 4.4, where some surface current data are presented from Pilot Charts for a tropic region.

At point A :  $\dfrac{\partial u}{\partial x} \simeq \dfrac{[(-0.25) - (-0.25)]\,ms^{-1}}{5 \times 10^5\ m} = 0$ .

At point B :  $\dfrac{\partial u}{\partial x} \simeq \dfrac{[(+0.25) - (+0.30)]\,ms^{-1}}{5 \times 10^5\ m} = -10 \times 10^{-8}\ s^{-1}$ .

Here we are approximating the derivatives by taking velocity differences over a finite (and by human size scale a very large) distance.  Such 'finite difference' approximations are good if the velocities vary in a smooth, nearly linear manner between the points where the velocities are measured.  The test of such methods is that they should give results consistent with observations; in the case of vertical velocities which are very hard to observe directly, because they are so small, such tests must be indirect.  We should also point out that in this particular example the available velocity information only consists of averages over approximately 5° squares at best, so finer scale estimates are not possible.

Therefore at the centre point E of the area, taking the mean of the values at A and B :

$\dfrac{\partial u}{\partial x} = -5 \times 10^{-8}\ s^{-1}$ .

Similarly, calculating $\partial v/\partial y$ at C and D and taking the mean for point E we get

$$\frac{\partial v}{\partial y} = -8.3 \times 10^{-8} \text{ s}^{-1} .$$

Then from         $\frac{\partial w}{\partial z} = -(\frac{\partial u}{\partial x} + \frac{\partial u}{\partial y})$ ,

we have           $\frac{\partial w}{\partial z} = -(-5 \times 10^{-8} - 8.3 \times 10^{-8}) \text{ s}^{-1}$ ,

$$= +13.3 \times 10^{-8} \text{ s}^{-1} .$$

Now $\partial w/\partial z$ is positive and w = 0 at the surface on the average. Therefore below the surface w must be negative (i.e., downward) since it must increase (become less negative) as the surface is approached where the value is to be zero. Thus $(\partial u/\partial x + \partial v/\partial y)$ being negative and hence $\partial w/\partial z$ being positive implies surface convergence at E.

Since w = 0 at the surface, where z = 0, the vertical velocity, $w_h$, at depth, h below the surface (i.e., at z = -h) is given by:

$$w_h = \int_0^{-h} dw = \int_0^{-h} \frac{\partial w}{\partial z} \cdot dz = -\int_0^{-h} \left( \frac{\partial u}{\partial x} + \frac{\partial v}{\partial y} \right) \cdot dz .$$

(A reader not familiar with integral notion is referred to Appendix I.)

In the numerical example above, if the convergence were constant from the sea surface to the bottom of the homogeneous layer, taken to be at 50 m depth, the vertical velocity at 50 m depth would be:

$$w_{50} = \int_0^{-50} (13.3 \times 10^{-8}) \cdot dz = -6.7 \times 10^{-6} \text{ m s}^{-1}$$

$$= 0.58 \text{ m day}^{-1} \text{ down} .$$

The fact that the layer is homogeneous and therefore presumably subject to considerable mixing (otherwise it would probably be stratified) provides some evidence that the convergence is uniform. With sufficient stirring, the horizontal velocities u and v should be independent of depth in the homogeneous layer. Then $[\partial u/\partial x + \partial v/\partial y]$ should also be independent of depth.

As the vertical velocities at shallower depths are correspondingly less, the time for a particle of water to sink from the surface to 50 m depth would be quite large. As $w = 13.3 \times 10^{-8} \cdot z$ and the time ($\delta t$) to sink $\delta z$ is $\delta t = \delta z/w$, then the time to sink from $z_1$ to $z_2$

$$= \int_{z_1}^{z_2} \frac{1}{w} \cdot dz$$

$$= 7.5 \times 10^6 \int_{z_1}^{z_2} \frac{1}{z} \cdot dz$$

$$= 7.5 \times 10^6 \ln(z_2/z_1) \text{ sec}$$

$$= 87 \ln(z_2/z_1) \text{ days} .$$

For example, the time to sink from 1 m depth to 50 m depth would be (87 · ln 50) = 340 days or almost one year!

It should be noted that this numerical example has been presented principally to illustrate the *use* of the equation of continuity and is for an open ocean situation. For regions of active upwelling, usually found along the east sides of the oceans, the vertical velocities may be greater - recent measurements indicate values of the order of $10^{-4} \, m \, s^{-1}$ or $10 \, m \, day^{-1}$ off the west coast of North America. Nevertheless these speeds are much less than the typical horizontal ones.

Continuity also shows why w << u or v. The ocean is very thin, the depth to width ratio being similar to that of a sheet of very thin paper! For uniform convergence, the vertical velocity at z = -H, $w_H \simeq H \cdot U/L$ where U is a change of horizontal velocity component over a distance L. Since $(H/L) = 0(10^{-3})$ or less for the whole ocean, then $w = 0(10^{-3} \times U)$ or less, i.e., of the order of one one-thousandth of the horizontal velocities or less.

Finally the effects neglected in deriving the equation of continuity (4.4) are very small compared with $(\partial u/\partial x + \partial v/\partial y)$. Thus the assumption of constant volume allows us to estimate $\partial w/\partial z$ and hence w; in fact, the errors in the estimation are much larger than the neglected effects.

# CHAPTER 5
# Stability and Double Diffusion

Here we consider whether or not the variation of density with depth in the
ocean is likely to cause the water to move vertically. If there is light
fluid on top of heavy fluid then there will be no tendency for motion to
occur. However, if there is heavy fluid above light fluid there will be a
tendency for the heavy fluid to sink and the light to rise - the density
distribution is unstable. Thus we must examine the vertical density gradient
to determine whether the fluid is stable, i.e., resists vertical motion, is
neutral, i.e., offers no resistance to vertical motion, or is unstable, i.e.,
tends to move vertically of its own accord. If $\partial \rho / \partial z < 0$ (density increases
with depth) we might expect the fluid to be statically stable so that if no
motion is occurring the density distribution will not cause motion to occur.
If $\partial \rho / \partial z > 0$, we expect the fluid to be unstable.

When considering the density distribution in relation to stability we cannot
ignore compressibility, i.e., the variation of density with pressure, which
means with depth. In the case of neutral stability, if a fluid parcel is
moved up or down adiabatically (that is, with no heat exchange with its
surroundings) and without salt exchange with its surroundings and then brought
to a stop it will not tend to move further because wherever it is moved to it
will have the same density as the surrounding fluid. Density must increase
with depth because a parcel which is moved down will be compressed but must
then have the same density as its surroundings. Thus in the neutral stability
case, $\partial \rho / \partial z < 0$ and the fluid appears to be quite stable if the pressure
effect is overlooked. At the same time, compression causes heating which
decreases density, although not nearly enough to overcome the increase of
density due to increased pressure and so, in the neutral stability case,
temperature increases with increased depth, assuming that salinity effects
can be ignored. If we neglect pressure effects we might think that the fluid
is slightly unstable because we have colder fluid above warmer fluid.

To take the compressibility into account one might try to consider the potential density or its anomaly, $\sigma_\theta$, defined in Chapter 2. It is the density of
the fluid when taken adiabatically to a reference pressure with the adiabatic
temperature change taken into account. This procedure can be used in the
atmosphere (provided no condensation or evaporation occurs). In fact, an
apparent or virtual potential temperature is used; this is the potential
temperature which dry air would have if it had the same potential density as
the moist air. Unfortunately, because of the complicated and non-linear
equation of state for sea water, using potential density to determine static
stability does not always work in the ocean. For example, North Atlantic
Deep Water has a slightly larger potential density than Antarctic Bottom Water.

23

Fig. 5.1.    Water properties for calculation of stability.

However, the former is found *above*, not below, the latter. The temperature
and salinity differences between these two water masses are sufficient that
the variation of compressibility with these parameters leads to the *in situ*
density (at the same depth) of the Antarctic water being slightly greater than
that of the North Atlantic Water, so that the Antarctic water flows under the
North Atlantic water.

The apparent instability in this case is caused by the fact that the reference
pressure for $\sigma_\theta$ is taken at the surface ($p = 0$). If a reference pressure close
to the *in situ* pressure is used then zero vertical variation of this potential
density will indicate neutral stability. However, to consider the stability
over the whole water column no single reference pressure is satisfactory in
all cases and it is necessary to calculate a local value of stability as a
function of depth as described below.

## Criterion for Static Stability (E)

Suppose that the density of a stationary water mass changes with depth in some
arbitrary manner and that at level 1 (Fig. 5.1), depth $= -z$  pressure $= p$, the
*in situ* water properties are ($\rho$, S, T). Then a parcel of water is moved a
short distance vertically from level 1 to level 2 without exchanging heat or
salt with its surroundings. At level 2, the depth $= -(z + \delta z)$ and pressure =
$p + \delta p$, and the surrounding water properties are ($\rho_2$, $S_2$, $T_2$). The water
properties of the parcel at level 2 will be ($\rho'$, S, $T + \delta T$) and its pressure =
$p + \delta p$. Here $\delta T$ is the adiabatic change of temperature due to change of

pressure, i.e., $\delta T = \left(\dfrac{dT}{dp}\right)_{adiabatic} \cdot \delta p$. As $\delta p = -\rho \cdot g \cdot \delta z$ (refer to

Appendix I) $\delta T = -\left(\dfrac{dT}{dp}\right)_{adiabatic} \cdot \rho \cdot g \cdot \delta z = -\Gamma \cdot \delta z$ where $\Gamma$ stands for the
adiabatic temperature gradient. It is the change in temperature with depth
caused by pressure change and is positive, i.e., compression causes the temperature to increase. At level 2 the restoring force on the parcel of volume

$\delta V_2$ will be:     F  =  buoyant upthrust - weight .

By Archimedes' Principle, buoyant upthrust = weight of surrounding fluid
displaced.   Hence:

$$F = \delta V_2 \cdot \rho_2 \cdot g - \delta V_2 \cdot \rho' \cdot g \tag{5.1}$$

$$= \delta V_2 \cdot g \cdot (\rho_2 - \rho')$$

and its acceleration if released will be:

$$a_z = \frac{F}{M} = \frac{\delta V_2 \cdot g (\rho_2 - \rho')}{\delta V_2 \cdot \rho'}$$

$$= \frac{g \cdot \left[ \rho + \left(\frac{\partial \rho}{\partial z} \cdot \delta z\right)_W - \rho - \left(\frac{\partial \rho}{\partial z} \cdot \delta z\right)_P \right]}{\rho \cdot \left[ 1 + \left( \frac{1}{\rho} \cdot \frac{\partial \rho}{\partial z} \cdot \delta z \right)_P \right]} \tag{5.2}$$

where the subscript W refers to the surrounding water and the subscript P to the parcel.

In equation 5.2 the change of density of the surrounding water

$$\left( \frac{\partial \rho}{\partial z} \cdot \delta z \right)_W = \left( \frac{\partial \rho}{\partial S} \cdot \frac{\partial S}{\partial z} + \frac{\partial \rho}{\partial T} \cdot \frac{\partial T}{\partial z} + \frac{\partial \rho}{\partial p} \cdot \frac{\partial p}{\partial z} \right)_W \cdot \delta z$$

and of the water parcel

$$\left( \frac{\partial \rho}{\partial z} \cdot \delta z \right)_P = \left( -\frac{\partial \rho}{\partial T} \cdot \Gamma + \frac{\partial \rho}{\partial p} \cdot \frac{\partial p}{\partial z} \right)_P \cdot \delta z$$

because the salinity does not change as it is measured in g/kg and is therefore independent of pressure effects.

Now $(dp/dz)_W = (dp/dz)_P$, and if the changes of salinity and temperature between levels 1 and 2 are not large, then $(\partial \rho / \partial p)_W = (\partial \rho / \partial p)_P$ because the $\delta S,_P$ and $\delta T,_P$ terms in the specific volume anomaly (equation 2.1) are small and slowly varying. Also, $(1/\rho) \cdot (\partial \rho / \partial z) \cdot \delta z$ in the denominator of equation 5.2 vanishes in the limit as $\delta z \to 0$ and may be neglected.

Then equation 5.2 becomes:

$$\frac{a_z}{g} = \frac{1}{\rho} \left[ \frac{\partial \rho}{\partial S} \cdot \frac{\partial S}{\partial z} + \frac{\partial \rho}{\partial T} \cdot \left( \frac{\partial T}{\partial z} + \Gamma \right) \right] \cdot \delta z \tag{5.3}$$

which is the ratio of the restoring acceleration of the displaced parcel to the acceleration due to gravity. Hesselberg defined the stability E of the water column as:

$$E = \left( -\frac{a_z}{g} \right) \quad \text{for} \quad \delta z = \text{unit length},$$

i.e., $$E = -\frac{1}{\rho} \left[ \frac{\partial \rho}{\partial S} \cdot \frac{\partial S}{\partial z} + \frac{\partial \rho}{\partial T} \cdot \left( \frac{\partial T}{\partial z} + \Gamma \right) \right] m^{-1} . \tag{5.4}$$

If E > 0, i.e., positive, the water is stable and a parcel displaced a short distance vertically will tend to return to its original position. Because it has inertia it will tend to overshoot its original position and then to oscillate about it, hence the stability of the water may be related

to the occurrence of internal waves (Chap. 12). If $E = 0$, the water is neu-
trally stable and a displaced parcel will tend to remain in its displaced
position. If $E < 0$, i.e., negative, the water will be unstable and a parcel
which is displaced will tend to continue its displacement, i.e., overturn of
the water should occur.

## Numerical Values for Stability

In the open ocean, values of E in the upper 1,000 m are of the order of
$100 \times 10^{-8} m^{-1}$ to $1,000 \times 10^{-8} m^{-1}$, the largest values generally occurring in
the upper few hundred metres. Below 1,000 m depth, values decrease to less
than $100 \times 10^{-8} m^{-1}$ and in deep trenches values close to $1 \times 10^{-8} m^{-1}$ are
found. In these latter cases, $\partial S/\partial z$ is generally very small so that its
effect on stability is negligible. Then as $E \to 0$ this means that $\partial T/\partial z \to -\Gamma$,
i.e., the temperature change with depth *in situ* is close to the adiabatic
rate due to change of pressure. The adiabatic rate increases from about
$0.14C°/1,000$ m at 5,000 m to $0.19C°/1,000$ m at 9,000 m depth, the temperature
changes being positive for increase of depth, i.e., the *in situ* temperature
increases with depth in deep trenches.

Note that in equation 5.4, $\partial \rho/\partial S$ and $\partial \rho/\partial T$ are taken holding the other variables fixed (T,p and S,p respectively) at the local *in situ* values. This
formula is not computationally very convenient because tables for density are
not commonly available - it is the specific volume that is normally tabulated.
To use such tables we use the fact that $\alpha = 1/\rho$ and hence $(1/\alpha) \cdot (\partial \alpha/\partial S) =$
$-(1/\rho) \cdot (\partial \rho/\partial S)$ and $(1/\alpha) \cdot (\partial \alpha/\partial T) = -(1/\rho) \cdot (\partial \rho/\partial T)$. Making use of the expansion of $\alpha$ of equation 2.1 (omitting the $\delta_{S,T,p}$ term which is negligible)
equation 5.4 becomes

$$E = \frac{1}{\alpha} \left( \frac{\partial \Delta_{S,T}}{\partial S} \cdot \frac{\partial S}{\partial z} + \frac{\partial \Delta_{S,T}}{\partial T} \cdot \frac{\partial T}{\partial z} + \frac{\partial \delta_{S,p}}{\partial S} \cdot \frac{\partial S}{\partial z} + \frac{\partial \delta_{T,p}}{\partial T} \cdot \frac{\partial T}{\partial z} + \Gamma \cdot \left( \frac{\partial \Delta_{S,T}}{\partial T} + \frac{\partial \delta_{T,p}}{\partial T} \right) \right). \quad (5.5)$$

The first two terms usually dominate and may be recognized as an expansion of
$(\partial \Delta_{S,T}/\partial z)$. The term involving $\Gamma$ is generally quite small and may be ignored
except in deep water where E is small. If $E = 0$, the neutral stability case,
omitting the $\Gamma$ term would give an apparent E of about $-2 \times 10^{-8} m^{-1}$ near the
surface and about $-4 \times 10^{-8} m^{-1}$ at great depth, so the water appears slightly
unstable if the adiabatic temperature change with pressure (i.e., depth) is
neglected as noted before. The importance of the other terms can be estimated
by comparing them with the first two. The first and third terms have
$(1/\alpha) \cdot (\partial S/\partial z)$ as a common factor so we need only compare coefficients of this
common factor. $(\partial \delta_{S,p}/\partial S)$ is of opposite sign to $(\partial \Delta_{S,T}/\partial S)$; its magnitude
is much smaller near the surface but increases to about 10% of $(\partial \Delta_{S,T}/\partial S)$ at
5,000 m depth and about 15% at 10,000 m depth. $(\partial \delta_{T,p}/\partial T)$ has the same sign
as $(\partial \Delta_{S,T}/\partial T)$; it also is relatively small near the surface but becomes
comparable at depths greater than about 2,000 m and may dominate at great
depth.

It is not easy to give a general rule about the relative importance of the
temperature and salinity terms. As a first approximation, the first two terms
of equation 5.5 may be used:

$$E \simeq \frac{1}{\alpha} \left( \frac{\partial \Delta_{S,T}}{\partial S} \cdot \frac{\partial S}{\partial z} + \frac{\partial \Delta_{S,T}}{\partial T} \cdot \frac{\partial T}{\partial z} \right) = \frac{1}{\alpha} \cdot \frac{\partial \Delta_{S,T}}{\partial z}. \quad (5.6)$$

If the calculated values of E are less than $50 \times 10^{-8} m^{-1}$ then the other terms
should be included.

The thermosteric anomaly $\Delta_{S,T}$ is normally calculated from $\sigma_t$ since they are directly related, as shown in Chap. 2, and $\sigma_t$ is usually calculated and tabulated along with S and T values during the first stage of data processing. Thus it is convenient to have an approximate formula for E in terms of $\sigma_t$. Suppose that we expand the *in situ* density in a manner similar to that used for $\alpha$ ( equation 2.1) as:

$$\rho = 1,000 + \sigma_t + \varepsilon_{S,p} + \varepsilon_{T,p} \qquad (5.7)$$

where a term of the form $\varepsilon_{S,T,p}$ has been omitted because it will be negligible. Substituting this expansion into 5.4 and using

$$\frac{\partial \sigma_t}{\partial S} \cdot \frac{\partial S}{\partial z} + \frac{\partial \sigma_t}{\partial T} \cdot \frac{\partial T}{\partial z} = \frac{\partial \sigma_t}{\partial z} \qquad \text{gives :}$$

$$E = -\frac{1}{\rho} \left( \frac{\partial \sigma_t}{\partial z} + \frac{\partial \varepsilon_{S,p}}{\partial S} \cdot \frac{\partial S}{\partial z} + \frac{\partial \varepsilon_{T,p}}{\partial T} \cdot \frac{\partial T}{\partial z} + \frac{\partial \rho}{\partial T} \cdot \Gamma \right) . \qquad (5.8)$$

The equivalent approximation to equation 5.6 is:

$$E \simeq -\frac{1}{\rho} \cdot \frac{\partial \sigma_t}{\partial z} . \qquad (5.9)$$

From equations 5.6 and 5.9 we see that a first approximation to stability is that $\Delta_{S,T}$ shall decrease with depth or that $\sigma_t$ shall increase with depth. Thus one can get an estimate of the sign of E just by looking at the tabulated values of $\Delta_{S,T}$ or $\sigma_t$. This is one of the reasons why $\sigma_t$ (or $\Delta_{S,T}$) is used rather than *in situ* values. (Another reason is that flow along constant $\sigma_t$ surfaces is easy since it is not restricted by static stability when equation 5.9 is a good approximation.) If one included the $\Gamma$ term of equation 5.8 in equation 5.9 then it would essentially be equivalent to $E = -(1/\rho) \cdot (\partial \sigma_\theta / \partial z)$. However, as we go a long way from the reference pressure ($p = 0$) the terms, other than the $\Gamma$ term, not in the approximate equations 5.6 and 5.9 become more important. Neglect of them leads to the apparent instability between the Antarctic Bottom Water and the North Atlantic Deep Water mentioned earlier.

Much of the effect of the pressure on the density cancelled out in deriving equation 5.4. Note also that the part of the pressure effect which cancelled is quite large. Suppose that we had just considered the gradient of *in situ* density. If the water were neutral, this gradient must be the same for both the water parcel and for the surrounding water:

$$\frac{1}{\rho} \cdot \left( \frac{\partial \rho}{\partial z} \right)_P = \frac{1}{\rho} \cdot \left( \frac{\partial \rho}{\partial p} \right)_{adiab} \cdot \frac{\partial p}{\partial z} = -g \cdot \left( \frac{\partial \rho}{\partial p} \right)_{adiab} ,$$

but     $\left( \frac{\partial \rho}{\partial p} \right)_{adiab} = \frac{1}{C^2}$     where C is the speed of sound ,

so     $-\frac{1}{\rho} \cdot \frac{\partial \rho}{\partial z} = \frac{g}{C^2} \simeq 400 \times 10^{-8} \text{ m}^{-1}$

and as stated earlier using the *in situ* density gives a false impression of quite stable conditions when the stability is actually neutral! If one wishes to use *in situ* density, $\rho_{S,T,p}$, then to correct for compressibility the stability is given by:

$$E = -\frac{1}{\rho} \cdot \frac{\partial \rho}{\partial z} - \frac{g}{c^2} .$$                                     (5.10)

(Again, in practice, one would usually have to use equation 2.1 to obtain $\alpha$ and then take $1/\alpha$ to get $\rho_{S,T,p}$.)

Although the water should be unstable and be expected to turn over whenever E is negative, in practice it is not uncommon to find values of E = -25 to $-50 \times 10^{-8}$ m$^{-1}$ in the upper 50 m of the sea with indications that the stratification is stable. As already shown, the neglect of the adiabatic temperature gradient and the other terms which are in equations 5.5 and 5.8 but not in 5.6 and 5.9 cannot account for such observations. It may be that some of these cases are in fact associated with weak convection but the observations are not detailed enough to detect it. Such apparent unstable situations may also be due to observational errors. In practice, E is calculated using finite differences with observations from discrete levels. The error in a $\sigma_t$ observation may easily be $5 \times 10^{-3}$ (see Chapter 2) and the error, $\Delta\sigma_t$, in the difference between two levels could easily be $10^{-2}$. With a depth difference $\Delta z$ of 20 m, the error in $E \simeq \pm(1/\rho) \cdot (\Delta\sigma_t/\Delta z)$ may be $\Delta E \simeq \pm 50 \times 10^{-8}$ m$^{-1}$. At greater depths where $\Delta z$ is larger (because the difference between observation levels is usually greater) the errors will be smaller, e.g., for $\Delta z = 500$ m and an error in $\Delta\sigma_t$ of $10^{-2}$, the error in E is only $2 \times 10^{-8}$ m$^{-1}$.

Tables of values of $\partial\rho/\partial S$, $\partial\rho/\partial T$ and $\Gamma$ (as $\partial\theta/\partial z$) for the calculations of E using equation 5.4 are given in Neumann and Pierson (1966) or one may use tables of $\Delta_{S,T}$, $\delta_{S,p}$, $\delta_{T,p}$ in equation 5.5.

The *Brunt-Väisälä frequency* N is given by:

$$N^2 = (g \cdot E) = g \cdot \left( -\frac{1}{\rho} \cdot \frac{\partial \rho_{T,S,p}}{\partial z} - \frac{g}{c^2} \right) \simeq g \cdot \left( -\frac{1}{\rho} \cdot \frac{\partial \sigma_t}{\partial z} \right) \text{(radians s}^{-1})^2.$$
                                                                                          (5.11)

The frequency in cycles sec$^{-1}$ (Hertz) is $N/2\pi = (g \cdot E)^{1/2}/2\pi$. It can be shown that this is the maximum frequency of internal waves in water of stability E. High values of N are usually found in the main pycnocline zone, i.e., where the vertical density gradient is greatest. This is usually in the thermocline in oceanic waters (where density variations are determined chiefly by temperature variations) or in the halocline in coastal waters (where density variations may be determined chiefly by salinity variations).


## DOUBLE DIFFUSION

Even though the water column may be statically stable at a particular time, instability may develop because sea water is a multi-component fluid and the rates at which heat and salt diffuse molecularly are different. A result is that if two water masses of the same density but different combinations of temperature and salinity are in contact, one above the other, the differential ('double') diffusion of these two properties may give rise to density changes which render the layers unstable. This is an active area of research and a review of the subject may be found in Turner (1973). The details are beyond the scope of the present book but the general ideas are interesting and double diffusion may play a significant role in small-scale mixing in the oceans and in the formation of 'fine' structure, the small-scale (one to a few metres)

variations in temperature and salinity gradients which have been found in the
oceans as observations have improved with the use of continuously recording
STD or CTD instruments (Salinity, Temperature, Depth, or Conductivity,
Temperature, Depth).

We consider the stability, starting with cases of positive static stability
but with no motion, because if there is motion, particularly turbulent motion
generated by velocity shear or strong static instability, turbulent diffusion
will dominate and probably prevent double diffusion effects from becoming
important. However, it seems that the ocean is sufficiently statically stable
in some parts that shear generated turbulence is suppressed and double
diffusive effects may be important.

Suppose that there is a layer of warmer, saltier water above cooler, fresher
water, such that the upper layer is of the same density or less dense than
the lower layer. Then the saltier water at the interface will lose heat to
the cooler water below faster than it will lose salt because the rate of
molecular diffusion of heat is about 100 times that of salt. If the density
difference between the layers is small, the saltier water above may become
heavier than the cooler, fresher layer below and sink downward into this
layer. Likewise the cold fresh water below the interface gains heat faster
than salt and may become light enough to rise into the upper layer. The
situation is referred to as one of 'double-diffusive instability'. The falling and rising motion occurs (in laboratory experiments) in the form of thin
columns and the phenomenon is called 'salt fingering'. There is evidence for
its occurrence in the ocean at the lower surface of the outflow of warm, saline
Mediterranean water from the Strait of Gibraltar into the cooler, fresher
Atlantic water.

If a layer of colder, fresher water is above a layer of warmer, saltier water,
the water just above the interface becomes lighter than that above it and
tends to rise while water below gets heavier and tends to sink. This phenomenon is called 'layering' and may lead to fairly homogeneous layers separated by thinner regions of high gradients of temperature and salinity. There
is evidence for its occurrence in the Arctic Ocean among other locations.

Both of these processes could lead to the vertical transports of heat and salt
being greater than the molecular diffusion rates, and to greater mixing than
would occur if these processes were not possible. Of course, once the motion
begins it may become dynamically unstable and break down into smaller scale
turbulent motions and become very complicated. Dynamic instability is
discussed briefly in the next section and also in Chapter 7.

The final possibility of a warmer, fresher layer above a cooler, saltier layer
does not allow a double-diffusive instability. The fresher water cools so
that it does not tend to rise but it cannot get colder than the saltier water
below and therefore it does not tend to sink. Similarly, the saltier water
does not tend to move up or down. For double diffusion to occur, the gradients of temperature and salinity across the interface must have the same
sign; then, since they affect density oppositely, double diffusion may occur.


## DYNAMIC STABILITY

Even if the water is statically stable and double diffusion is not permitted
by the temperature and salinity distributions, if motion is initiated it may

be dynamically unstable and it may break down into smaller-sized irregular turbulent motions. This possibility will be discussed further after we have examined the equations of motion.

Turbulent flows are familiar to everyone although they may not normally be labelled as such. Examples are the flow in most rivers, the gusty wind and the flow of water out of a tap, among many others. All these flows are very irregular both as a function of time at a fixed point and from point to point at a given time. The strong mixing caused by turbulent flow is often used, e.g., in stirring milk and sugar into coffee. After the stirring is stopped, the flow will gradually become more regular providing an example of non-turbulent flow, a type of flow which is less familiar in everyday experience.

# CHAPTER 6
# The Equation of Motion in Oceanography

Here we consider how Newton's Second Law of Motion ($F = m \cdot a$) can be written in a form which can be applied in oceanography. (Underlining a symbol, e.g., $F$, indicates that it is a vector quantity. Further discussion is given in Appendix I.)

This relation says that if a resultant force $F$ acts on a body of mass m, the body will acquire an acceleration or rate of change of velocity, $a$. Notice the adjective 'resultant', which means that more than one force may be acting simultaneously and we must first find the resultant of these, i.e., the net force, by appropriate vector addition; the acceleration will then be in the direction of this resultant force. In practice, vector equations such as $F = m \cdot a$ are usually broken down into three component equations so that the sum of the x-components of the forces equals the product of the mass times the x-component of the acceleration, etc.

The relation implies that if $F = 0$, then $a = 0$, i.e., there will be no *change* of motion but there may be persistent motion. This situation is governed by Newton's First Law of Motion, a special case of the Second Law.

Also, if we *observe* that $a = 0$, we can conclude that the resultant $F = 0$. In principle, this conclusion could mean that no forces at all were acting but in practice on earth this situation never occurs. For instance, there is always weight acting and often a reacting force balancing it, and if there is motion there is generally friction acting. When unaccelerated motion occurs, if we have information about some of the forces acting we can often learn something about the other forces. Notice also that $a = 0$ implies motion in a straight line; whenever we observe that the motion is curved there is a centripetal acceleration (see Appendix I) and therefore a resultant force must be acting.

It is convenient to write $a = F/m$ and think of the Law as stating that the observed acceleration is due to the resultant force acting per unit mass. In words we write :

Acceleration = (pressure + gravity + frictional + tidal force)/unit mass.

To make physical-mathematical deductions from this statement of a physical law we must first write mathematical statements of the forms of the forces; then we can try to 'obtain solutions' to the equations as explained below.

In vector form the equation is:

$$\frac{d\underline{V}}{dt} = -\alpha \cdot \nabla p \quad - \quad 2\underline{\Omega} \times \underline{V} \quad + \quad \underline{g} \quad + \quad \underline{F} \; . \hspace{2cm} (6.1)$$
$$\qquad\qquad\uparrow\qquad\qquad\uparrow\qquad\qquad\uparrow\qquad\uparrow$$
$$\qquad\text{Pressure}\quad\text{Coriolis}\quad\text{Gravity}\;\; \text{Other forces (all per unit mass)}$$

(We will explain shortly how we arrived at this equation and the meaning of
the symbols $\nabla$ and $\underline{x}$. $\underline{V}$ is the total velocity.)

This equation can be written as three component equations with the coordinates
x, y and z and their respective velocity components u, v and w being positive
in the east, north and upward directions respectively and the origin of
coordinates being at the sea surface:

|        | Pressure | Coriolis | Gravity | Other forces/unit mass |
|--------|----------|----------|---------|------------------------|
| (x) | $\dfrac{du}{dt} = -\alpha \cdot \dfrac{\partial p}{\partial x}$ | $+\;\; 2\Omega \cdot \sin\phi \cdot v \;\; - \;\; 2\Omega \cdot \cos\phi \cdot w$ | | $+ \qquad F_x$ |
| (y) | $\dfrac{dv}{dt} = -\alpha \cdot \dfrac{\partial p}{\partial y}$ | $-\;\; 2\Omega \cdot \sin\phi \cdot u$ | | $+ \qquad F_y \quad (6.2)$ |
| (z) | $\dfrac{dw}{dt} = -\alpha \cdot \dfrac{\partial p}{\partial z}$ | $+\;\; 2\Omega \cdot \cos\phi \cdot u$ | $- \quad g$ | $+ \qquad F_z \; .$ |

These equations 6.1 or 6.2 are called the equation(s) of motion; they are also
referred to as the equations of conservation of linear momentum. Similar
equations may be written for conservation of angular momentum but we will use
a related quantity called 'vorticity' in this book, as is customary in fluid
mechanics, and use of the word 'momentum' will imply linear momentum unless
otherwise stated.

## OBTAINING SOLUTIONS TO THE EQUATIONS, INCLUDING BOUNDARY CONDITIONS

In these equations, the quantities u, v and w are the components of the velocity of the water and they describe the 'motion of the ocean' - they are
what the physical oceanographer, particularly the dynamic oceanographer, wants
to learn about. Together with the pressure p, they form the four unknowns in
the equations. If we add the equation of continuity (4.4) we have four
equations and four unknowns. The other quantities assumed to be known are:
x, y, z for position, $\alpha$ = specific volume (from the pressure (or depth) and
the observed distributions of temperature and salinity), $\Omega$ = the angular
velocity of rotation of the earth, $\phi$ = geographic latitude (from y), while
$\underline{F}$ and its components $F_x$, etc., represent frictional and tidal forces which we
will introduce later. We shall consider the more complicated situation when
S, T and $\alpha$ are also taken to be unknown in Chapter 10 when discussing thermohaline effects.

Then 'obtaining solutions' to the equations of motion means finding (or guess-
ing) values for u, v and w, in terms of the known quantities, which 'satisfy
the equations'. This statement means that if one substitutes these values
for u, v and w in the equations, these will balance, i.e., the numerical
value of the time differential of u (i.e., du/dt) on the left of the first of
equations 6.2 must be identical with that of the right hand side when numerical values for the quantities are substituted there. And similarly for the

other two equations.  Note that the same expressions for u, etc., must be
used in all equations where they occur.

It must also be noted that the expressions for u, v and w must simultaneously
satisfy two other conditions:

   (1)   the equation of continuity $\frac{\partial u}{\partial x} + \frac{\partial v}{\partial y} + \frac{\partial w}{\partial z} = 0$ ,

and (2)   the boundary conditions.

With regard to item (1), the assumption of incompressibility leading to this
simple form of continuity eliminates acoustic or sound waves from the
possible solutions to equations 6.2 because such waves depend for their
existence on the medium being compressible.  We shall not discuss sound waves
explicitly in this book although many of the properties of waves given in
Chap. 12 are also applicable to them.

Item (2) above means that the velocity components u, v and w must behave in
a reasonable manner at the boundaries of the ocean, i.e., at the bottom, shore
and air/sea surface.  For instance, if x occurs in the expression (solution)
for u, it must be in such a way that u becomes zero at a north-south shore
and w must become zero at the ocean bottom if it is level.  More generally,
there can be no flow through the boundaries, so the component of flow normal
(i.e., perpendicular) to the boundary must vanish.  Next to solid boundaries
the component of flow along the boundary (the 'tangential' component) must
vanish too, i.e., there must be 'no slip' at solid boundaries.  This condition
is based on the observed behaviour of almost all real fluids and is a consequence of the molecular nature of materials and the surface interactions
between the solid and fluid.

Sometimes it may be possible to relax the 'no slip' condition.  The velocity
along the boundary may decrease to zero at the boundary in a relatively thin
'boundary layer'.  One may be able to find a solution for the interior region
of the flow which does not satisfy the no slip condition and a 'boundary layer
solution' which goes from the interior solution to no slip in the thin layer.
If one is only interested in the interior then one can consider the interior
solution alone and use a 'free slip' boundary condition to find it.  Caution
is required, however, because the assumption that the effects of the boundary
do not penetrate into the region of interest may be incorrect and erroneous
solutions may result.

One procedure for obtaining a solution is simple (even crude) - invent one
and then see if it will satisfy all the conditions; then examine it, using
observed information, to see if it describes a likely motion of the water.
We can ease the task by simplifying the equations by ignoring the F terms and
by ignoring the acceleration terms.  A number of useful solutions for this
case are known and will be discussed in Chapter 8.

The procedure becomes more difficult when expressions for fluid friction are
introduced as F terms, because we are still uncertain about the physical
details of turbulent fluid friction except in a few special cases.  Finding
solutions is even more difficult when the acceleration terms are included.
The equations then become non-linear and very difficult to deal with mathematically if we need analytic solutions, i.e., algebraic expressions for the
velocities.  (The alternative procedure of solving the equations numerically
will be discussed in Chapter 11.)

Fig. 6.1.    For the derivation of the pressure term in the equation of
             motion.


'*Non-linear*' means that the unknowns occur in combination in the equations,
e.g., as $v \cdot (\partial u / \partial y)$, and non-linear equations seem to be impossible to solve
in general.    (One may be able to show that unique solutions should exist but
have no general way of finding them.)    The non-linearity of the equations and
the presence of turbulence are not unrelated as we shall discuss in the next
chapter.


## THE DERIVATION OF THE TERMS IN THE EQUATION OF MOTION

### The Pressure Term

Imagine a rectangular volume, in a fluid, of sides $\delta x$, $\delta y$ and $\delta z$, fixed in a
coordinate system itself fixed relative to the solid earth (Fig. 6.1).    Then
the force in the x-direction on this volume due to the hydrostatic pressure
will be $+p \cdot \delta y \cdot \delta z$ on the left face and $-(p + \delta p) \cdot \delta y \cdot \delta z$ on the right face,
where the minus sign indicates that it acts in the negative x-direction.    The
net 'pressure force' in the x-direction is the sum of these two or
$-i \cdot \delta p \cdot \delta y \cdot \delta z = -i \cdot (\partial p / \partial x) \cdot \delta x \cdot \delta y \cdot \delta z$, where the unit vector i denotes
the x-direction.*    The force per unit volume is $-i \cdot (\partial p / \partial x)$, and the force
per unit mass $= -i \cdot (\partial p / \partial x) \cdot (1/\rho) = -i \cdot \alpha \cdot (\partial p / \partial x)$.    Then considering all
directions, the total pressure force/unit mass will be:

$$ - \alpha \cdot \left( \underline{i} \cdot \frac{\partial p}{\partial x} + \underline{j} \cdot \frac{\partial p}{\partial y} + \underline{k} \cdot \frac{\partial p}{\partial z} \right) \; = \; - \alpha \cdot \nabla p \; . $$

---

*Here, as usual, we have omitted terms of higher order in $\delta x$ which vanish in
the limit as $\delta x$ becomes small.    From now on, this approach will be used where
appropriate without stating so explicitly every time.

Here, $\nabla$ is short for $(i \cdot \partial/\partial x + j \cdot \partial/\partial y + k \cdot \partial/\partial z)$, called the 'gradient
operator' and $j$ and $k$ are unit vectors denoting the y and z directions
respectively. The gradient of a quantity (e.g., $\nabla p$) is always a vector and
therefore it is not necessary to underline it as we have done with other
vectors, such as F or V. The minus sign indicates that if p increases to the
right, then the pressure force acts to the left. (Derivatives of quantities,
in this case p, in a particular direction, e.g., $\partial p/\partial x$, are often called
gradients too, as a convenient term for the component in a particular direction of the total gradient.)

## Transforming from Axes Fixed in Space to Axes Fixed in the Rotating Earth

The Coriolis term arises because we normally make observations relative to
axes fixed to the earth which is itself rotating about its axis. The equation
of motion F = m · a however applies only when a is measured relative to axes
'fixed in space', (i.e., in what is called an inertial coordinate system
which is one whose origin is not accelerating). For practical purposes this
is a system 'fixed relative to the distant stars'. Obviously it is more
convenient for the oceanographer to make his measurements relative to points
and directions on earth and so the equation of motion must be adjusted to suit
this rotating frame of reference.

A mathematically straightforward and exact transformation from ideal axes
fixed in space to practical rotating earth axes (e.g., Neumann and Pierson,
Lacombe or Batchelor texts in the Further Reading list) gives in vector
form:

$$\underline{a}_f \;=\; \left(\frac{dV'}{dt}\right)_f \;=\; \left(\frac{dV}{dt}\right)_e + 2\underline{\Omega} \times \underline{V} + \underline{\Omega} \times (\underline{\Omega} \times \underline{R}) \qquad (6.3)$$

where the subscript f means relative to fixed axes and the subscript e means
relative to the earth. On the right hand side, the first term is the acceleration relative to axes fixed to the earth, the second term is the Coriolis
acceleration and the third term is the centripetal acceleration required to
make an object on the earth's surface rotate with the earth. The other
symbols are: V' = velocity relative to fixed axes, V = velocity relative to
the earth, R = the vector distance of the body from the centre of the earth,
and $\Omega$ = angular velocity of rotation of the earth. Its value is $2\pi$ radians
in one sidereal day or $7.29 \times 10^{-5}$ rad s$^{-1}$. (One sidereal day (23 h 56 min 4 s,
or 86164 s) is the time required for the earth to rotate once about its axis,
relative to the fixed stars. Since the earth revolves about the sun it must
turn a little further to point back to the sun and complete one solar day −
hence the solar day is a little longer than the sidereal day.) The 'x' in a
term such as $2\Omega \times V$ represents what is called a vector product. The reader
unfamiliar with this vector operation need not be concerned because we will
write down and use the components of this operation in the component
equations.

The equation of motion relative to fixed axes is:

$$\left(\frac{dV'}{dt}\right)_f \;=\; -\,\alpha \cdot \nabla p + \underline{g}_f + \underline{F} \;. \qquad (6.4)$$

When transformed to earth axes using 6.3 we get:

$$\left(\frac{dV}{dt}\right)_e \;=\; -\,\alpha \cdot \nabla p - 2\underline{\Omega} \times \underline{V} + \underline{g}_f - \underline{\Omega} \times (\underline{\Omega} \times \underline{R}) + \underline{F} \;. \qquad (6.5)$$

In this equation, the term on the left is the acceleration relative to the earth and the terms on the right are the forces per unit mass acting, i.e., the accelerations due to these forces. The transformation simply adds two apparent forces/unit mass, i.e., $-2\underline{\Omega} \times \underline{V}$, termed the Coriolis force, and $-\underline{\Omega} \times (\underline{\Omega} \times R)$, the negative of the centripetal acceleration (which is some- times called the centrifugal acceleration (force/unit mass), see Appendix I). The true forces in equation 6.4 are unchanged.

## Gravitation and Gravity

*Gravitation* is the name given to the attractive force between masses, recog- nized first by Newton. Its magnitude is expressed by $F_g = G \cdot (M_1 \cdot M_2)/r^2$ where $M_1$ and $M_2$ are the sizes of two masses and r is the distance between their centres. It is an attractive force acting along the line connecting the centres of the masses. (This expression is only true for two masses whose sizes are small compared with r or for two spheres whose density distri- bution is radially symmetrical. These conditions are sufficiently well satisfied in the case of the earth and a small object on it, and for the earth and moon when we consider tidal theory.) G is the Gravitational Constant. The gravitational force provides the $g_f$ in the absolute equation of motion (6.4). In the relative equation (6.5), the term $\underline{\Omega} \times (\underline{\Omega} \times R)$ is the centripetal acceleration required to make a body at a distance $\underline{R}$ from the centre of the earth circulate about the earth's axis with angular velocity $\underline{\Omega}$. As usual for bodies in contact with the earth it is provided by a portion of the gravitational acceleration $g_f$, as shown in Fig. 6.2. (The maximum value of the magnitude of the centripetal acceleration is only about 0.3% of the gravitational acceleration.) The difference, $[g_f - \underline{\Omega} \times (\underline{\Omega} \times R)]$, is referred to as the *acceleration due to gravity*, i.e., it is the familiar acceleration g of a body falling freely near the earth (in the absence of friction). In future we combine $[g_f - \underline{\Omega} \times (\underline{\Omega} \times R)]$ as g. At the surface of the earth it depends only on geographical position. It is a maximum at the poles (where the needed centripetal acceleration vanishes and $g_f$ is also a maximum because the polar radius is slightly less than the equatorial radius) and is a minimum at the equator (where the needed centripetal acceleration is a maximum and $g_f$ is a minimum). However, as the variation of g from pole to equator is only about 0.5% we will neglect it, and also will neglect the very small variation with depth below the ocean surface, and will take the value of g as constant at 9.80 m s$^{-2}$.

Notice that in the component equations 6.2 the acceleration due to gravity occurs only in the z-component equation because the z axis is, by definition, taken parallel to the local direction of this acceleration. It appears with a minus sign because the acceleration is down while the positive direction for z is taken as up.

## The Coriolis Terms

The terms containing $\underline{\Omega}$ in equation 6.1 and $\Omega$ in equation 6.2 and the $(\underline{\Omega} \times V)$ term in equation 6.5 are called the 'Coriolis' acceleration terms (named after G. Coriolis, 1835, although they had been recognized by others before him). As will be seen from equations 6.2 there are four terms. Of these, the component $2\Omega \cdot \cos \phi \cdot w$ in the x-equation is very small compared with the other terms in that component equation, because w is so small, and this component of the Coriolis acceleration will be neglected. In addition, the Coriolis term $2\Omega \cdot \cos \phi \cdot u$ in the z-equation is small compared with the pressure term and with g, but is not necessarily small compared with their difference which

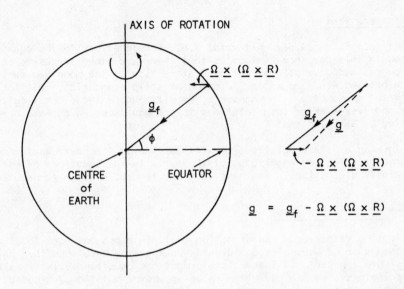

Fig. 6.2.    Showing how the gravitational acceleration ($g_f$) is reduced to the
acceleration due to gravity ($g$) in providing the centripetal
acceleration ($\underline{\Omega} \times (\underline{\Omega} \times \underline{R})$) required.   Note that the size of the
centripetal acceleration is exaggerated.    Its magnitude is
$\Omega^2 \cdot R \cdot \cos \phi$ and it acts perpendicular to the axis of rotation.

is itself usually small in the sea.   However, this z-component Coriolis term
is usually neglected in dynamic oceanography.

Only two Coriolis terms are left and these are in the x- and y-component
equations, which depend only on the horizontal components of velocity.   These
two terms can be combined as a horizontal Coriolis acceleration $\underline{C}_H = 2\Omega \cdot \sin \phi \cdot$
$\underline{V}_H \times \underline{k}$ where $\underline{V}_H = \underline{i} \cdot u + \underline{j} \cdot v$ = the horizontal component of the total
velocity.   The direction of $\underline{C}_H$ must be perpendicular to both $\underline{k}$ (the unit
vector in the vertically upward direction) and to $\underline{V}_H$, i.e., it is horizontal
and directed at right angles to and to the right of $\underline{V}_H$ in the northern
hemisphere, to the left in the southern hemisphere.

The factor $2\Omega \cdot \sin \phi$ is often abbreviated to f so that $2\Omega \cdot \sin \phi \cdot u = f \cdot u$,
etc.

The magnitude of $\underline{C}_H$ for a current speed of $1\,m\,s^{-1}$ (or approximately 2 knots)
which is fairly typical for major ocean currents is: at $\phi = 90°$ (pole),
$C_H = 1.5 \times 10^{-4}\,m\,s^{-2}$; at $\phi = 45°$, $C_H = 1 \times 10^{-4}\,m\,s^{-2}$ and at $\phi = 0°$ (equator),
$C_H = 0$.   These are small accelerations; an acceleration of $10^{-4}\,m\,s^{-2}$ would
take about 40 hours to give a body, starting from rest, a speed of $14\,m\,s^{-1}$
($\simeq 50\,km\,h^{-1}$ or 30 miles per hour)!   Another way to put it is to say that a
body lying on a frictionless slope of $10^{-4}/9.8 \simeq 1$ in $10^5$ or $1\,cm$ drop in
$1\,km$ horizontally would experience the acceleration of $10^{-4}\,m\,s^{-2}$.   This slope

is of the same magnitude as is calculated for the mean slopes of the sea
(neglecting wave slopes).

## Other Accelerations

The final term $\underline{F}$ (force per unit mass, i.e., acceleration, in the equations
of motion in this chapter) reminds us that there are other forces to be taken
into account, such as the gravitational attraction of the moon and sun,
friction between wind and water, friction at solid boundaries, friction within the water mass, etc.   In Chapter 8, we will ignore these complicating
factors and just examine some solutions to the equations which have been
simplified by assuming that $\underline{F} = 0$ .

Before looking at these solutions we will examine in the next chapter some of
the characteristics and magnitudes of the terms in the equations of motion,
after briefly discussing in the next section, the coordinates to be used.


## COORDINATE SYSTEMS

In writing the vector equation of motion (6.1) in component form (6.2) we
used *rectangular* or *Cartesian* coordinates because the equations then have a
fairly simple expression.  The vector form of the equation (6.1) is valid in
any coordinate system, an advantage in using vector notation.  (Indeed, one
may derive an equation in component form in one system, e.g., our derivation
of the pressure force in rectangular coordinates, transform it to vector form
and expect it to be valid in any coordinate system when properly transformed.)

If we are considering the motion of the ocean over the whole earth, then rectangular coordinates are clearly not appropriate.  Spherical coordinates must
be used.  The earth is not exactly spherical but is approximately elliptical
in cross-section from north to south pole with an ellipticity of about 1/300.
(This shape or 'figure' of the earth is a consequence of the fact that the
gravitational acceleration is partially used to provide the required centripetal acceleration.)  However, the error involved in using the equations in
spherical form is only about 0.5% and can be neglected.

In this book, we shall write the component equations in rectangular form.
The equations are simplest in this form so that it is easiest to illustrate
the principles with them.  Also, for many phenomena they are a consistent
approximation.  If the horizontal area being considered is not too large then
we can work on a plane tangent to the sphere and use a rectangular system
with negligible errors.   For phenomena of relatively small scale, e.g.,
100 km or so, this tangent plane is called the  *f-plane*  because for such
small north-south distances the Coriolis parameter, f, may be taken to be
constant at the value at the centre of the area.  (In the Arctic Ocean, where
f is near to its maximum and varying only slowly, this approximation of using
rectangular (or cylindrical) coordinates with f = constant may be used for
many phenomena over rather larger regions.)  For relatively larger areas,
with $\phi$ varying over a few tens of degrees, between mid-latitudes and the
equator, the tangent plane approximation is called the  *β(beta)-plane* .  Here,
while a rectangular coordinate approximation is used, the variation of f with
latitude is taken as $f = (f_0 + \beta \cdot y)$ where $f_0$ is the value  of f at the mid-
latitude of the region and $\beta = \partial f/\partial y$ is given the value at the mid-latitude of
the area.  The quantity $\beta$ is the variation of the Coriolis parameter with
latitude.

# CHAPTER 7

# The Role of the Non-Linear Terms
# in the Equations of Motion

Before discussing some special cases of the application of the equation of motion, we will examine in this chapter the role of the non-linear terms in the equation and will make some estimates of their quantitative significance.

## THE NON-LINEAR TERMS IN THE EQUATION OF MOTION

### The Friction Term for the Instantaneous Velocity

Consider the equation of motion (equation 6.2) for the x-component:

$$\frac{du}{dt} = -\alpha \cdot \frac{\partial p}{\partial x} + 2\Omega \cdot \sin\phi \cdot v - 2\Omega \cdot \cos\phi \cdot w + \text{friction} + \text{tidal forces.} \tag{7.1}$$

The tidal force terms can be written down using Newton's Law of Gravitation and can be taken to be known, although when this equation was being examined in the early days of fluid mechanics attention was directed chiefly to laboratory flows where such terms are not important. We shall omit the tidal terms until we consider their effects in Chapter 13. In order to get a system which can be solved, an expression for the friction is needed. One can consider a small element fixed in space and consider the forces on it associated with the molecular nature of the fluid and differences in velocity within the fluid. Based on observational evidence, Newton hypothesized and it was later verified, using non-turbulent flows, that the frictional forces were related to spatial derivatives of velocity (e.g., $\partial^2 u/\partial y^2$) multiplied by a viscosity coefficient which is a property of the fluid. Thus the frictional effects could be expressed in terms of the velocity, and a closed system of equations could be obtained. By this statement we mean that the number of equations equals the number of unknowns and, at least in principle, they can be solved. The friction term in equation 7.1 takes the form

$$\nu \cdot \left( \frac{\partial^2 u}{\partial x^2} + \frac{\partial^2 u}{\partial y^2} + \frac{\partial^2 u}{\partial z^2} \right)$$

where $\nu$ is the kinematic molecular viscosity, $\nu = \nu(S,T,p)$. A typical value for water is $10^{-6} \text{ m}^2 \text{ s}^{-1}$ with a range of 0.8 to 1.8 times this value. In the derivation of this expression it has been assumed that the fluid is incompressible and terms of the form $(\partial\nu/\partial x) \cdot (\partial u/\partial x)$ have been neglected because they are small compared with those retained in realistic oceanographic cases. The derivation of the friction term in this form was done by Navier and Stokes and the equations of motion including it are called the 'Navier-Stokes equations'. (Details of the derivation which is mathematically straight-

forward, although the algebra may be complicated depending on the notation
used, may be found in more advanced texts or in fluid mechanics texts, e.g.,
Batchelor, 1967.)  We shall carry out a derivation of the $\nu \cdot (\partial^2 u/\partial z^2)$ term
later when discussing the wind-driven circulation in Chapter 9.  A term representing the resistance, due to molecular viscosity, to compression has been
omitted.  It is not important for the solutions which we shall consider but
viscosity does lead to damping of sound waves and may need to be retained
when studying acoustics in the sea.

Equation 7.1 applies to the instantaneous velocity of the fluid and is an
excellent approximation to describe the behaviour of fluids such as water,
i.e., it agrees with all the experimental results within measurement error.
However, real oceanic and atmospheric fluid motions may be turbulent (very
irregular in space and time) so that it may not be practical to solve the
equations exactly for them.  (For example, the details of boundary conditions
and the initial state of the fluid are never well enough known.)

What is the Source of the Difficulty?

The term du/dt applies to the acceleration of a piece of the fluid.  The terms
on the right hand side are written in Eulerian form (see Appendix I).  To use
the equation we must write du/dt in Eulerian terms.  (We could attempt to
write the right hand side in Lagrangian terms but it is more difficult to do
so and the same problem arises as in the Eulerian approach.)  In writing
du/dt in Eulerian terms (velocities at fixed points as functions of time) we
must take account of the fact that a particle of fluid, when it moves to
another point, must arrive at that point at a later time with the velocity
appropriate to the new point and time.

The form which the acceleration term takes is then the *'total'* or *'individual'*
derivative discussed in Appendix I for precisely this purpose and equation
7.1 becomes:

$$\frac{du}{dt} = \frac{\partial u}{\partial t} + u \cdot \frac{\partial u}{\partial x} + v \cdot \frac{\partial u}{\partial y} + w \cdot \frac{\partial u}{\partial z} = \text{(terms on the right} \qquad (7.2)$$
$$\text{as in equation 7.1)}$$

local rate      advective rates of
of change       change due to
due to          motion
time
variation

The advective terms are called 'non-linear' because the velocities occur as
squares (e.g., $u \cdot (\partial u/\partial x) = (1/2) \cdot [\partial (u^2)/\partial x]$)or as products between different
velocity components and their derivatives (e.g., $v \cdot (\partial u/\partial x)$.  Because of these
non-linear terms a small perturbation (variation) may grow into a large
fluctuation - these terms can cause instability and lie behind the presence
of the turbulence which occurs whenever they are sufficiently large compared
with the frictional terms which tend to remove velocity differences.

Scaling and the Reynolds Number

To estimate what is meant by 'large' in this case, let us consider the ratio
$(u \cdot \partial u/\partial x)/(\nu \cdot \partial^2 u/\partial x^2)$ of one of the non-linear terms to one of the molecular
friction terms.  If we take both u and $\partial u$ to be of order U (a typical velocity

magnitude) and $\partial x$ of order L (a typical distance over which the velocity varies by U), then the ratio above is of the order $(U^2/L)/(\nu \cdot U/L^2) = U \cdot L/\nu$ which is called the *Reynolds Number* (Re) for a fluid flow. It is a measure of the ratio of the non-linear, also called inertial, terms to frictional terms in the equation of motion.

This process of *'scaling'* or *'ordering'* terms is very often used in fluid mechanics because we cannot solve the full equations. By doing this scaling we may find that some terms may be neglected and thus simplify the analysis. We shall look at all of the terms presently but here we want to use the Reynolds Number example to explain in detail the assumptions involved and the limitations of the approach. As a specific example consider flow in a pipe, the problem on which Osborne Reynolds worked. Here the radius of the pipe provides a length scale, L, for variations of velocity which goes from zero at the pipe wall to a maximum at the centre. The flow rate at the centre, U, provides a velocity magnitude. Now to estimate the size of the non-linear terms (velocity times velocity gradient) we use $U^2/L$. The non-linear terms will not be of exactly this value and may vary in size from one part of the flow to another but they should be proportional to this quantity with a con-stant of proportionality of order 1 (i.e., between 0.1 and 10) if we have chosen our scales properly. Likewise, $\nu \cdot (\partial^2 u/\partial x^2)$ etc. should be proportion-al to $\nu U/L^2$. Hence the ratio, the Reynolds Number, gives an estimate of the relative importance of the two terms. This is another example of a finite difference approximation - in this case a rather crude one.

This dimensionless number, Re, is a very important quantity in determining the character of the flow. Indeed it can be shown that for a uniform density fluid with no rotation the solution to the equations is completely determined by the geometry of the boundaries and the Reynolds Number (although we may not be able to find the solution in a particular case).* Flows which have the same geometry and Reynolds Number are said to be 'dynamically similar', that is when scaled properly, the flows are identical. As a specific example, if we double the pipe diameter and adjust the total volume flow per unit time so that the velocities are half as large, and $U \cdot L/\nu$ remains the same, the two flows will look the same. Thus dynamical similarity can be used to organize one's experimental efforts. For example, if we are trying to determine under what conditions the flow in the pipe will change from smooth, laminar flow to irregular, turbulent flow, the principle of dynamic similarity tells us that the transition will, for a given geometry, occur at some particular value of Re. Thus to investigate the change we need to vary the flow rate only and not the pipe size or viscosity which are more difficult. The flow in the pipe will not be turbulent if Re < 1,000. If the entrance to the pipe is smoothly flared and care is taken to reduce velocity fluctuations before the fluid enters the pipe it is possible to maintain laminar flow for Re up to about 100,000. This variation in the value of Re at transition is caused by changes in geometry, in this case the entrance conditions. The fact that transition to turbulence does not occur until Re $\simeq$ 1,000 shows that the non-linear terms do not become of dominant importance until this value is reached. In the pipe flow they are zero until transition to turbulence occurs but in many flows (e.g., around objects in the fluid) they usually begin to modify the solution (from the one obtained by assuming that they are zero) when

---

*
If a free surface, i.e., a non-solid boundary, is present, the dimensionless parameter $U^2/g \cdot L$, called the Froude Number, must be considered too.

Re $\simeq$ 1 , although turbulence will not be produced until Re becomes much larger.  Once Re becomes larger than $10^5$ to $10^6$, depending on geometry, turbulence is very likely to occur unless there is some stabilising influence such as density stratification as we shall discuss later in the chapter.

If we use the Gulf Stream as an oceanographic example, U $\sim$ 1 m s$^{-1}$ and L $\sim$ 100 km = $10^5$ m and $\nu \sim 10^{-6}$ m$^2$ s$^{-1}$, so that Re $\sim 10^{11}$ and the flow will definitely be turbulent.

We conclude from this example that the non-linear effects are very strong compared with the molecular friction effects.  We can, in fact, ignore molecular friction in the open sea; it only becomes important very close to solid boundaries and in removing energy from turbulent flow at small scales to prevent it from growing without limit, i.e., molecular friction is important only for low values of Re which occur at low values of U and/or of L.

## Reynolds' Stresses

Although molecular friction may be neglected in most aspects of the dynamics of ocean motions, it must not be assumed that there are no forces opposing the motions or giving rise to redistribution of energy and other properties. When the motion is turbulent, so that it includes rapidly fluctuating components in addition to any mean flow, then the non-linear terms give rise to terms in the equations of motion which have the physical character of friction and they, and similar terms in the heat and salt conservation equations (discussed in Chapter 10), give rise to more rapid distribution of momentum, heat and salt than would occur with purely molecular processes.  These are the so-called *'Reynolds stresses'* (forces/unit area) and *'fluxes'* (transports/unit area) which appear in the equations for the mean or average motion of a turbulent fluid (the Reynolds equations, named after Osborne Reynolds who first derived them and the equations for heat and salt conservation).

## EQUATIONS FOR THE MEAN OR AVERAGE MOTION

Because of the nature of turbulent flow it does not appear to be practical to solve for the detailed velocities so let us examine the possibility of writing equations for the average motion.  The average used will be taken to be a *time average* over a suitable period (which might be a few minutes up to several months, depending on the phenomenon).  Following Osborne Reynolds, who first suggested the approach, the variables u, v, w and p are split into a *mean* and a *fluctuating* part, e.g., u = $\bar{u}$ + u' where the overbar denotes an average and $\overline{u'}$ = 0 by definition; $\bar{u}$ + u' , etc., are then substituted into equation 7.2 and the average is taken.

Consider first the average of $\partial u / \partial t$ :

$$\overline{\frac{\partial u}{\partial t}} = \frac{1}{T} \int_0^T \frac{\partial u}{\partial t} \cdot dt = \frac{[u(T) - u(0)]}{T} \text{ where T is the averaging period.}$$

Now u must have some upper limit because the available energy sources are limited and frictional losses always occur (and usually increase as u increases).  Thus as T gets large, this term will become negligible.

Fig. 7.1.   Time variation of some property P at a point.  The solid curve is
            the total quantity P(t) as a function of time t, the dashed curve is
            a  possible mean  $\bar{P}$ , while the difference between the two at any
            instant would be taken to be the fluctuating part P'.


In practice we might wish to consider separation into a time-varying mean and
fluctuations about it, if we were interested in variations due to tides or in
seasonal changes.  The sketch of Fig. 7.1 illustrates how we might do so.
(Note that the separation is not always as straightforward as shown in the
figure but such problems do not affect the general results of this section; a
more detailed discussion is left for more advanced texts.)  Thus we retain
$\partial\bar{u}/\partial t$  in the average of equation 7.2 so that later we can consider how
important this term is in the equation for the mean flow.  For the other terms
the order of averaging and differentiating may be interchanged if required.
First let us consider the terms on the right-hand side of equation 7.2.

The pressure term becomes:

$$\overline{(\bar{\alpha} + \alpha') \cdot \frac{\partial(\bar{p} + p')}{\partial x}} \;=\; \overline{\bar{\alpha} \cdot \frac{\partial\bar{p}}{\partial x}} + \overline{\bar{\alpha} \cdot \frac{\partial p'}{\partial x}} + \overline{\alpha' \cdot \frac{\partial\bar{p}}{\partial x}} + \overline{\alpha' \cdot \frac{\partial p'}{\partial x}} \;.$$

In the second term on the right, $\bar{\alpha}$ is already an average and does not change
in the averaging process, so this term is $\bar{\alpha} \cdot (\overline{\partial p'}/\partial x) = 0$ because $\bar{p'}$ $(=0)$ is
independent of x .  Likewise, the third term vanishes.  In general, any
average of a term which contains a single fluctuating quantity will vanish.
The term $\alpha' \cdot (\partial p'/\partial x)$ may not vanish if fluctuations in $\alpha'$ and p' are related
[e.g., $\alpha'$ might be either of the same or of the opposite sign to $(\partial p'/\partial x)$ on
average (although not always)].  However, the $\alpha$ variations in the ocean are
very small compared with $\bar{\alpha}$, and $\partial p'/\partial x$ will be of the order $\partial\bar{p}/\partial x$ or less, so
$\alpha' \cdot (\partial p'/\partial x)$ is negligible compared with $\bar{\alpha} \cdot (\partial\bar{p}/\partial x)$.

The first Coriolis term $\overline{2\Omega \cdot \sin\phi \cdot (\bar{v} + v')} = 2\Omega \cdot \sin\phi \cdot (\bar{v} + \overline{v'})$ because $\Omega \cdot \sin\phi$
is a constant at a particular location and the average of a sum is the sum of
the averages.  Now $\overline{v'} = 0$ , so this Coriolis term becomes $2\Omega \cdot \sin\phi \cdot \bar{v}$ ; likewise the other Coriolis term becomes $2\Omega \cdot \cos\phi \cdot \bar{w}$ when averaged.

In the frictional term, for example:

$$\nu \, \frac{\partial^2 (\bar{u} + \overline{u'})}{\partial x^2} \;=\; \nu \cdot \left( \frac{\partial^2 \bar{u}}{\partial x^2} + \frac{\partial^2 \overline{u'}}{\partial x^2} \right) = \nu \, \frac{\partial^2 \bar{u}}{\partial x^2} \quad \text{because } \overline{u'} = 0 \text{ everywhere}$$

and therefore its spatial derivatives must vanish.

Thus all the terms on the right-hand side of equation 7.2 are of the same form for both the total flow and for the mean flow, which is also true for the corresponding terms in the y and z equations.

Now let us examine the advective acceleration terms on the left-hand side of equation 7.2 which become:

$$(\bar{u} + u') \cdot \frac{\partial (\bar{u} + u')}{\partial x} + (\bar{v} + v') \cdot \frac{\partial (\bar{u} + u')}{\partial y} + (\bar{w} + w') \cdot \frac{\partial (\bar{u} + u')}{\partial z} \;.$$

When we average them any term which contains a single fluctuating quantity will vanish on average, as we showed above. Hence the average becomes

$$\left( \bar{u} \cdot \frac{\partial \bar{u}}{\partial x} + \bar{v} \cdot \frac{\partial \bar{u}}{\partial y} + \bar{w} \cdot \frac{\partial \bar{u}}{\partial z} \right) + \left( \overline{u' \cdot \frac{\partial u'}{\partial x}} + \overline{v' \cdot \frac{\partial u'}{\partial y}} + \overline{w' \cdot \frac{\partial u'}{\partial z}} \right).$$

The first three terms combined with $\partial \bar{u}/\partial t$ give $d\bar{u}/dt$, the total or individual derivative using the mean rather than the total velocity, leaving the other three terms involving the fluctuating components.

Hence, if we collect everything together the *Reynolds equation* for $\bar{u}$ has the same form as 7.2, with mean quantities used for total quantities, plus the three new terms above involving velocity fluctuations. These new terms must represent the effect of the velocity fluctuations or 'turbulence' on the mean motion. Note that they arise from the non-linear terms in the Navier-Stokes equation; the non-linear nature of the equations and the possible existence of turbulence and its possible frictional effects on the mean flow are not un-related, as noted before. Note too that equations 6.1 and 6.2, because we did not write a specific form for the friction, may be taken to be applicable to either the total or mean velocities.

We are still faced with the problem of writing down specific expressions for the friction in order to produce a closed (i.e., complete) set of equations. The approach of Navier and Stokes produces a set which is closed but which cannot be applied to high Reynolds Number turbulent oceanic flows in practice. Reynolds' approach shows how the non-linear terms give rise to turbulence effects on the mean flow and gives explicit expressions for these effects in terms of the velocity fluctuations. However, the system is still not closed because we have now added three more unknowns, u', v', and w'. In principle one might attempt to observe these turbulence terms. Such observations can be made, with great difficulty, at a single location but to do so in the detail required for a large region is simply not practical. To improve our under-standing of turbulence effects and to be able to 'parameterize' these effects in terms of mean flow quantities and their gradients (i.e., space derivatives) many more observations will be needed. (By *'parameterize'* we mean write down expressions for the turbulence quantities in terms of quantities which we can observe more easily or calculate from our equations, in this case the mean

velocities and their gradients and the density distribution or perhaps the
static stability based on it.)  This problem of 'closure' - completing the
equation set - remains as a fundamental problem in studying turbulent flows
and remains unsolved in any general way.  It is possible to work out from the
Navier-Stokes equation an equation describing the behaviour of such terms as
$u'u'$, $u'v'$, $u'w'$, etc., from which the turbulence terms above can be calculated.
However, these equations contain terms involving averages of triple products
of fluctuating quantities which also come from the non-linear terms.  An
equation for these can be obtained but it involves quadruple products and so
on *ad infinitum*.  There are always more unknowns than equations.  One must
use observational knowledge and physical intuition to provide the necessary
additional equations.  The test of any closure scheme is that predictions
using it agree with observations.  In the following we shall outline the
simplest closure scheme - an analogy with molecular friction effects.

## Reynolds Stresses and Eddy Viscosity

The analogy which will be discussed, of introducing an 'eddy' or 'turbulent'
viscosity of much greater magnitude than the molecular one, does not produce
very exact results except in special cases.  However, if we can show using
this analogy that turbulent friction effects are small we can perhaps 'solve'
the equations ignoring friction and expect realistic results.  Note also that
in the mean flow or Reynolds equation the non-linear terms are not likely to
be dominant.  The 'breakdown' into turbulence will mix momentum and reduce
the spatial derivatives of the mean flow to the point where the non-linear
terms based on the mean flow do not dominate, i.e., are not the largest terms.
Thus the mean flow equations are likely to be, at worst, weakly non-linear.
Provided that we can overcome the closure problem in a reasonable way we have a
good chance of solving the equations, by numerical calculations on a computer
if necessary (as described in Chapter II) although many analytical methods
(i.e., writing down appropriate mathematical formulae) are available for
weakly non-linear equations.

First we shall rewrite the turbulence terms using the equation of continuity
of volume for an incompressible fluid, $\partial u/\partial x + \partial v/\partial y + \partial w/\partial z = 0$ .  If we
average this equation then $\partial \bar{u}/\partial x + \partial \bar{v}/\partial y + \partial \bar{w}/\partial z = 0$ ; subtracting this
average equation from the original equation then $\partial u'/\partial x + \partial v'/\partial y + \partial w'/\partial z = 0$
or $\nabla \cdot V' = 0$ in mathematical shorthand.  Thus the total velocity, the mean
velocity and the fluctuating velocity all satisfy continuity of volume.  To
the turbulence terms of the Reynolds equation we add $u' \cdot (\nabla \cdot V')$ which is zero
and so does not change the value, only the mathematical form.  For the x
component:

$$u' \cdot \frac{\partial u'}{\partial x} + v' \cdot \frac{\partial u'}{\partial y} + w' \cdot \frac{\partial u'}{\partial z} + u' \cdot \left( \frac{\partial u'}{\partial x} + \frac{\partial v'}{\partial y} + \frac{\partial w'}{\partial z} \right) = \frac{\partial}{\partial x} u'u' + \frac{\partial}{\partial y} u'v' + \frac{\partial}{\partial z} u'w'$$

and the equation for $\bar{u}$ is:

$$\frac{d\bar{u}}{dt} = - \bar{\alpha} \cdot \frac{\partial \bar{p}}{\partial x} + 2\Omega \cdot \sin\phi \cdot \bar{v} - 2\Omega \cdot \cos\phi \cdot \bar{w} + \nu \cdot \left( \frac{\partial^2 \bar{u}}{\partial x^2} + \frac{\partial^2 \bar{u}}{\partial y^2} + \frac{\partial^2 \bar{u}}{\partial z^2} \right)$$

$$- \frac{\partial}{\partial x} \overline{u'u'} - \frac{\partial}{\partial y} \overline{u'v'} - \frac{\partial}{\partial z} \overline{u'w'} .$$

$$(7.3)$$

This is the Reynolds equation for the x component of velocity.

Now a term such as $\nu \cdot (\partial^2\bar{u}/\partial x^2)$ can be written as $\partial(\nu \cdot \partial\bar{u}/\partial x)/\partial x$ (the form which this term had before it was assumed that $\nu$ is essentially uniform as shown later in Chapter 9), and $\rho \cdot \nu \cdot (\partial\bar{u}/\partial x)$ is the stress (force/unit area) in the x-direction due to molecular effects and a gradient of $\bar{u}$ in the x-direction. We can therefore identify $-\rho \cdot \overline{u'u'}$ as a stress due to the turbulence. It is the derivatives of these stresses which produce net forces on a small volume of fluid (just as does the derivative of the pressure as shown in Chapter 6). The stress mechanisms are qualitatively similar - the molecular effect is produced by molecules bouncing back and forth and exchanging momentum, the turbulent stress is producec by 'chunks' of fluid moving back and forth and exchanging momentum with the surrounding fluid. The latter is more effective because the distance moved and the mass involved are *much* larger. Stresses such as $-\rho \cdot \overline{u'u'}$, $-\rho \cdot \overline{u'v'}$, $-\rho \cdot \overline{u'w'}$ (and the other averaged quadratic products of u', v' and w') are termed Reynolds stresses, again after Osborne Reynolds who first derived them. By analogy with the molecular case we might suppose that these stresses are related to the mean velocity gradients by some sort of 'viscosity' (an *'eddy'* or *'turbulent'* viscosity),

$$\text{e.g.,} \quad -\overline{u'u'} = A_x \cdot \frac{\partial\bar{u}}{\partial x} \; ; \quad -\overline{u'v'} = A_y \cdot \frac{\partial\bar{u}}{\partial y} \; ; \quad -\overline{u'w'} = A_z \cdot \frac{\partial\bar{u}}{\partial z} \; . \quad (7.4)$$

Unlike the molecular case we use different values of eddy viscosity ($A_x$ etc.) for different directions, since they may be different (particularly between the vertical and horizontal directions because of static stability).*

Then a term such as $-\frac{\partial}{\partial x}\overline{u'u'}$ becomes $\frac{\partial}{\partial x}(A_x \cdot \frac{\partial\bar{u}}{\partial x})$. It is quite common to take $A_x$ outside the derivative, either based on the argument (again in analogy with molecular viscosity) that terms such as $\frac{\partial A_x}{\partial x} \cdot \frac{\partial\bar{u}}{\partial x}$ are less important, or that the analogy is crude (which it is) and this further assumption is no worse than the initial one (which may or may not be true, depending on the case). With this final neglect of space variations of the A's relative to the other terms, the turbulent friction terms become, in the x-direction:

$$A_x \cdot \frac{\partial^2\bar{u}}{\partial x^2} + A_y \cdot \frac{\partial^2\bar{u}}{\partial y^2} + A_z \cdot \frac{\partial^2\bar{u}}{\partial z^2} \qquad\qquad (7.5)$$

where the A's are called 'eddy' viscosities. Note that they are, like $\nu$, kinematic (dimensions $[L^2T^{-1}]$ with units $m^2 s^{-1}$) and the terms in expression 7.5 have dimensions of force/unit mass, i.e., acceleration. One must multiply the A's by $\rho$ to get dynamic viscosity which when multiplied by $\partial^2\bar{u}/\partial x^2$ gives a force acting on unit volume. In the CGS system of units, where $\rho \sim 1\ gm\,cm^{-3}$, the dynamic and kinematic viscosities have about the same numerical value but in SI with $\rho$ of the order of $1,000\ kg\,m^{-3}$ they do not. [In the literature, the symbol A (or other symbol for viscosity) has sometimes been used for kinematic and sometimes for dynamic viscosity, so some care is required when extracting numerical values. In this book, kinematic viscosity will be used throughout.]

---

*This is the simplest way to define the eddy viscosity. However, this definition does not preserve the symmetry of the Reynolds' stresses, e.g., $-\overline{v'u'} = A_x \cdot (\partial\bar{v}/\partial x)$ is not necessarily the same as $-\overline{u'v'} = A_y \cdot (\partial\bar{u}/\partial y)$ although it should be. Usually either $\partial\bar{v}/\partial x$ or $\partial\bar{u}/\partial y$ will dominate and one can pick from the flow which is appropriate. It is possible to get around this problem but equation 7.4 (and 7.5 following) are sufficient for our purposes and we shall leave further discussion of this point for more advanced texts.

Unlike coefficients of molecular viscosity, the eddy viscosity coefficients
are not constant for a particular fluid and temperature but vary from place
to place and from time to time, and with the particular motion involved.  They
are *not properties of the fluid but of the flow!*  Values are up to $10^{11}$ times
those for kinematic molecular viscosity.  Many attempts have been made to
express $A_x$, etc., in terms of the mean velocities and their derivatives but
no generally applicable results have been obtained.  We must therefore remember
that the eddy viscosity terms in the above form are just in interim measure to
represent one of the effects of turbulence until we understand this feature of
fluid motion well enough to represent it more exactly.

The eddy viscosity approach does give good results in some cases, e.g., in the
atmospheric surface layer, the first few tens of metres above the surface.
In this layer $A_z$ varies linearly with z and the solution of the equations
(which are the same as for the ocean) using the eddy viscosity form for the
friction term agrees very well with observations.  Presumably the flow near
the ocean bottom could be treated in the same way but observations of the flow
in this part of the ocean are quite limited.

When we introduce the eddy viscosity (including the molecular viscosity in
it), the equations of motion for the x and y components are:

$$\frac{du}{dt} = \frac{\partial u}{\partial t} + u \cdot \frac{\partial u}{\partial x} + v \cdot \frac{\partial u}{\partial y} + w \cdot \frac{\partial u}{\partial z} =$$

$$= - \alpha \cdot \frac{\partial p}{\partial x} + f \cdot v - 2\Omega \cdot \cos \phi \cdot w + A_x \cdot \frac{\partial^2 u}{\partial x^2} + A_y \cdot \frac{\partial^2 u}{\partial y^2} + A_z \cdot \frac{\partial^2 u}{\partial z^2}$$

$$2\Omega \sin\theta \qquad\qquad\qquad\qquad\qquad\qquad\qquad\qquad\qquad (7.6)$$

$$\frac{dv}{dt} = \frac{\partial v}{\partial t} + u \cdot \frac{\partial v}{\partial x} + v \cdot \frac{\partial v}{\partial y} + w \cdot \frac{\partial v}{\partial z} =$$

$$- \alpha \cdot \frac{\partial p}{\partial y} - f \cdot u \qquad\qquad + A_x \cdot \frac{\partial^2 v}{\partial x^2} + A_y \cdot \frac{\partial^2 v}{\partial y^2} + A_z \cdot \frac{\partial^2 v}{\partial z^2}$$

where u, v, w, $\alpha$, and p are average quantities, the overbar having been omitted
for simplicity.  (Unless otherwise stated we assume that we are discussing the
mean motion equations from now on.)

## SCALING THE EQUATIONS OF MOTION

The equations of motion in the form of equations 7.6 and the corresponding z
component equation (given shortly) are complicated and non-linear (although
usually only weakly) and are generally not solvable explicitly.  Before giving
up in mathematical despair, let us examine the various terms to make rough
estimates of their size - it may be possible initially to neglect some of them
but still leave equations which refer to physical reality in the ocean and
describe actual motions, even if only approximately.  Later, we can reintroduce
some of the neglected terms and obtain more exact mathematical descriptions of
the motion.

What we will do is refer to the data bank of descriptive oceanography to find
out what may be the sizes of the various terms so that we can decide which
are the most important in particular situations.

First let us consider the main body of the oceans away from strong currents (such as the Gulf Stream or Kuroshio) and away from the sea surface where the frictional influence of the wind is important. We can return to these regions later.

The Pacific Ocean is roughly 12,000 km across and the Atlantic 6,000 km, so let us take a horizontal length scale (L) of 1,000 km = $10^6$ m as typical of large-scale features of the ocean circulation. Typical horizontal speeds (U) are of the order of 0.1 m s$^{-1}$. We will take a vertical scale length (H) of $10^3$ m, a reasonable fraction of the total depth (world ocean average = 4,000 m).

First we will estimate a typical vertical speed (W) using the equation of continuity:

$$\frac{\partial w}{\partial z} = - \left( \frac{\partial u}{\partial x} + \frac{\partial v}{\partial y} \right)$$

for which the order of magnitude of the terms is:

$$\frac{W}{H} \simeq \frac{U}{L}$$

$$W \simeq \frac{U \cdot H}{L} \simeq \frac{10^{-1} \cdot 10^3}{10^6} \simeq 10^{-4} \, m\,s^{-1} \quad .$$

For a typical time scale (T) we take 10 days $\simeq 10^6$ s, considering shorter periods to be turbulent components for the moment. For the Coriolis acceleration, at latitude $\phi = 45°$ then $2\Omega \cdot \sin 45° \simeq 2 \times 7.3 \times 10^{-5} \times 0.71 \simeq 10^{-4} s^{-1}$, while $g \simeq 10$ m s$^{-2}$. For the pressure term, $\alpha \simeq 10^{-3} m^3 \, kg^{-1}$ and $p \simeq 10^4$ kPa $= 10^7$ Pa for $z = -10^3$ m from the hydrostatic equation.

Values estimated for $A_x$ and $A_y$ vary from 10 to $10^5 \, m^2 \, s^{-1}$ and we will use $10^5 \, m^2 \, s^{-1}$. For $A_z$ estimates range from $10^{-5}$ to $10^{-1} \, m^2 \, s^{-1}$ and we will use $10^{-1} \, m^2 \, s^{-1}$. By using maximum values for $A_x$, $A_y$, $A_z$, we should get upper limits for the size of the friction terms.

We see that these estimates for eddy viscosity vary widely. Part of this variation is due to the fact that they are properties of the flow, not of the fluid, and part is due to the way in which they are obtained. For example, by measuring or estimating the other important terms in the equations one may obtain the friction term by difference and then calculate the eddy viscosity from $A_x =$ friction / $(\partial^2 u/\partial x^2)$. Alternatively, one may adjust the eddy viscosities in a solution (either analytical or numerical) to make it fit the observations as well as possible. A simple rough approach (probably good within a factor of 100) is to use the fact that the non-linear terms are about the same size as the turbulent friction terms which we pointed out in the previous section. Then:

$$\frac{U^2}{L} \simeq A_x \cdot \frac{U}{L^2} \simeq A_y \cdot \frac{U}{L^2} \simeq A_z \cdot \frac{U}{H^2} \quad \text{or} \quad A_x \simeq U \cdot L \quad \text{and} \quad A_z \simeq \frac{H^2}{L^2} \cdot A_x \quad .$$

Thus with H/L $\simeq 10^{-3}$, $A_z \simeq 10^{-6} \cdot A_x$ as for the estimates given. The fact that $A_z \ll A_x$ or $A_y$ is due to the static stability caused by stratification which both inhibits vertical turbulent transfers and forces the flow to be nearly horizontal. (Indeed the generally stable stratification in the upper part of the sea is essential in making typical vertical velocities $W \ll U$, and leading to circulations with $H \ll L$.) Note that $A_x \simeq U \cdot L$ is equivalent to saying that a Reynolds Number based on eddy viscosity is of order 1. Using $U = 0.1\,\mathrm{m\,s^{-1}}$ and $L = 10^6\,\mathrm{m}$ gives $A_x$ and $A_y \simeq 10^5\,\mathrm{m^2\,s^{-1}}$, at the upper end of the range of estimated values. Lower values may occur because the flows on which they are based are of smaller scale (smaller L or U) or have an eddy viscosity Reynolds Number > 1.

The vertical component equation of motion corresponding to equation 7.6 is:

$$\frac{dw}{dt} = \frac{\partial w}{\partial t} + u \cdot \frac{\partial w}{\partial x} + v \cdot \frac{\partial w}{\partial y} + w \cdot \frac{\partial w}{\partial z} =$$

$$- \alpha \cdot \frac{\partial p}{\partial z} + 2\Omega \cdot \cos \phi \cdot u - g + A_x \cdot \frac{\partial^2 w}{\partial x^2} + A_y \cdot \frac{\partial^2 w}{\partial y^2} + A_z \cdot \frac{\partial^2 w}{\partial z^2} \qquad (7.7)$$

and will have scale sizes as follows:

$$\frac{W}{T} + \frac{U \cdot W}{L} + \frac{V \cdot W}{L} + \frac{W^2}{H} =$$

$$\alpha \cdot \frac{10^7}{H} + 2\Omega \cdot \cos \phi \cdot U - g + 10^5 \cdot \frac{W}{L^2} + 10^5 \cdot \frac{W}{L^2} + 10^{-1} \cdot \frac{W}{H^2}$$

i.e., $\qquad 10^{-10} + 10^{-11} + 10^{-11} + 10^{-11} =$

$$+ \quad 10 \quad + \quad 10^{-5} \quad - 10 + 10^{-11} + 10^{-11} + 10^{-11}.$$

In this equation, all the terms are very much smaller than the pressure term and g and so we can ignore all except these two and will be left with the *hydrostatic equation* (derived in Appendix I), i.e., $\alpha \cdot (\partial p/\partial z) = -g$ (7.8) correct to about 1 part in 1 million, even when the water is moving with typical open ocean speeds and even though we have chosen values for the eddy viscosities at the high end of the observed range in order to have the frictional forces at the high end of their range for the open ocean. (It is left as an exercise for the reader to show that the hydrostatic equation still applies even in faster currents such as the Gulf Stream where the maximum speed is about $3\,\mathrm{m\,s^{-1}}$ and the stream width is $\sim 100$ km.)

Note that the non-linear terms are all of the same size as a result of our estimating the vertical speed W from the horizontal speeds using the equation of continuity. Also the friction terms are all of the same (small) size as a result of our choice of the H, L, $A_x$ and $A_z$ values. This result will hold also for the other component equations and therefore when examining them we will need to look at only one non-linear and one friction term to estimate their size.

Looking now at one of the horizontal component equations:

$$\frac{\partial u}{\partial t} + u \cdot \frac{\partial u}{\partial x} + \cdots = - \alpha \cdot \frac{\partial p}{\partial x} + f \cdot v - 2\Omega \cdot \cos \phi \cdot w + A_x \cdot \frac{\partial^2 u}{\partial x^2} + \cdots$$

The order of magnitude of the terms is:

$$\frac{U}{T} + \frac{U^2}{L} + \cdots = - \alpha \cdot \frac{\partial p}{\partial x} + f \cdot U - 10^{-4} \cdot W + 10 \cdot \frac{U}{L^2} + \cdots$$

or    $10^{-7} + 10^{-8} + \cdot \cdot \cdot = ? + 10^{-5} - 10^{-8} + 10^{-8} + \cdot \cdot \cdot$

or relatively:

   $10^{-2} + 10^{-3} + \cdot \cdot \cdot = ? + 1 \quad - 10^{-3} + 10^{-3} + \cdot \cdot \cdot$

The pressure term has been represented by a query here because we do not have direct measurements of $\partial p/\partial x$. We see, however, that it must be of the same size as the Coriolis term ($f \cdot v$) in order to balance the equation. Of the remaining terms the local acceleration term $\partial u/\partial t$ is the largest but even it is only about 1% of the Coriolis term for typical times of the order of 10 days and will be smaller for longer times. The second Coriolis term ($2\Omega \cdot \cos \phi \cdot w$) is small because of the typically small value of w. The non-linear terms for the mean motion are negligibly small and so are the friction terms in the interior of the water mass. Therefore, to an order of accuracy of 1% we have:

$$0 = - \alpha \cdot \frac{\partial p}{\partial x} + f \cdot v$$

$$0 = - \alpha \cdot \frac{\partial p}{\partial y} - f \cdot u \qquad \text{for the interior of the ocean.} \qquad (7.9)$$

$$0 = - \alpha \cdot \frac{\partial p}{\partial z} - g$$

These equations describe the relationships between the horizontal pressure distributions and the horizontal velocity components in the ocean, and the distribution of pressure as a function of depth and density ($\rho$) distribution ($\alpha = 1/\rho$) which is a function of the distribution of salinity, temperature and pressure. In principle, if we observe the distribution of salinity and temperature as a function of depth in the ocean we can calculate p from the z equation 7.9 and use it to find u and v from the x and y equations. Alternatively, for theoretical studies we could express the temperature and salinity distributions mathematically as functions of x, y and z, introduce the equation of state $\alpha = \alpha(T,S,p)$ from laboratory studies of the properties of sea water and heat and salt conservation equations, and solve the set of simultaneous equations (seven in all), an approach which we shall discuss in Chapter 10.

It appears, therefore, that the interior region of the ocean is described by a simple set of equations which can be solved, because non-linear effects are negligible. However, these simple equations do not give us a complete description because the boundary conditions for the interior of the ocean depend on the surface layers where wind friction acts, and on the lateral boundary layers (e.g., the Gulf Stream) where the dynamics are more complicated. A complete solution for the interior requires solutions for the outer regions, so that the problem is not fully solved. We can, however, ignore the boundary condition problem for a while and make use of the simple equations to find out quite a lot about the motion in the interior.

In the next chapter we shall look at a simple case where there are no true forces - just acceleration provided by the Coriolis effect. In this case we will be looking at a phenomenon of smaller linear scale because if:

   Acceleration term  =  Coriolis acceleration
then
   $\frac{U^2}{L} \simeq f \cdot U$   or   $L \simeq \frac{U}{f}$   and for   $U = 0.1 \, m \, s^{-1}$, $f \simeq 10^{-4}$,

then $L \simeq 10^3$ m or 1 km. In this example we have used a possible balance between terms to determine what the length scale must be, another way to use the scaling approach.

In the scaling of the large-scale flow in the interior we found that both non-linear and friction effects were very small. In other regions they may be more important. We found the Coriolis term to dominate - it turns out to be important for almost all large-scale flow phenomena.

To help to classify flow types in other regions it is useful to consider the ratios between non-linear and Coriolis terms and between friction and Coriolis terms:

$$\frac{\text{Non-linear term}}{\text{Coriolis term}} = \frac{U^2}{L} \cdot \frac{1}{f_0 \cdot U} = \frac{U}{f_0 \cdot L} = R_0 \quad .$$

Here $f_0$ is a typical value of $f$ for the region being considered and the ratio $R_0$ is called the *Rossby Number*. The second non-dimensional ratio is:

$$\frac{\text{Friction term}}{\text{Coriolis term}} = A_x \cdot \frac{U}{L^2} \cdot \frac{1}{f_0 \cdot U} = \frac{A_x}{f_0 \cdot L^2} = E_x \quad ,$$

$$\text{or} \quad = \frac{A_y}{f_0 \cdot L^2} = E_y \quad , \quad \text{or} \quad = \frac{A_z}{f_0 \cdot H^2} = E_z \quad .$$

These E's are called *Ekman Numbers*, e.g., $E_z$ is the vertical Ekman Number because it depends on the friction term involving spatial derivatives with respect to the vertical coordinate, often termed for brevity vertical friction. Likewise $E_x$ and $E_y$ are horizontal Ekman numbers. In the interior, $E_x \sim E_y$ and the symbol $E_H$ is often used. For the interior $R_0 \leq 10^{-3}$, $E_z \sim E_H \leq 10^{-3}$. In other regions they may not be so small but for the large scale circulation, values of the order of 1 are an upper limit.


## DYNAMIC STABILITY

What determines when a flow will become unstable so that it may break down into irregular small-scale motions leading to friction effects which are much larger than those due to the molecular nature of the fluid? As we have already noted, such effects seem to occur in the ocean because the apparent friction effects, as quantified by the eddy viscosities, are much larger than molecular ones. The horizontal eddy viscosity values are $10^7$ to $10^{11}$ times molecular values while the vertical values are 10 to $10^5$ times molecular values.

First consider a fluid which is not rotating so that Coriolis terms can be ignored. Also take the fluid to have constant and uniform density throughout, so that derivatives of density with respect to space coordinates vanish everywhere. This is an idealized example requiring a truly incompressible fluid and is often used without further explanation. It is quite easy, however, to construct a realistic example with the same properties. Take the salinity and potential temperature to be constant throughout. Then the static stability is neutral - there is no buoyant resistance to vertical motion because a displaced parcel always has the same density as the surrounding water. Alternatively, one can say that there are now no buoyancy effects - displaced parcels are never lighter or heavier than their surroundings. The salinity is uniform throughout but temperature and density both increase with depth and are not uniform. However, this real fluid case will behave exactly as the ideal constant density, truly incompressible, case so the results of the ideal case are of practical value.

Now for this simple situation it is the ratio of the non-linear terms to the molecular friction term, i.e., the Reynolds Number, which determines the *dynamic stability*. If Re > $10^6$, turbulent flow is likely. Suppose U = 0.01 m $s^{-1}$, a rather small velocity, then taking $\nu \simeq 10^{-6}\,m^2\,s^{-1}$, the characteristic length to make Re = $10^6$ is L = 100 m. As this length is rather small compared with the size of the ocean basins it would seem that turbulent flow is likely to occur everywhere. However, even in this simple case a large value of Re is not sufficient for turbulence to occur. In order that small velocity variations (also called 'perturbations') can grow they must have an energy source. It turns out that there is no energy source unless there are gradients in the flow. Thus if the flow is very uniform in velocity there is no energy source and molecular viscosity will smooth out the perturbations. Of course, the ocean is finite and near solid boundaries the velocity vanishes, leading to gradients if there is any flow at all and therefore turbulence will probably be present there. At the surface the wind acts, leading to velocity gradients and turbulence.

Another possibility is that although the Reynolds Number is large and velocity gradients are present, for a particular type of flow the non-linear terms remain small and the breakdown to turbulence does not occur. Surface waves provide such a case. Although Re may easily be $10^7$ or more they are weakly non-linear and not turbulent until wave breaking occurs. While the non-linear effects are small for surface waves they are not entirely negligible, as we shall discuss briefly in the chapter on waves.

## The Effect of Density Variations on Dynamic Stability

When density variations occur in the fluid they may enhance or diminish the mechanical effects. The static stability gives a measure of the effect. If it is negative (unstable) the vertical component of velocity fluctuations is enhanced. If it is positive (stable) the vertical component is diminished. If the turbulence persists it will tend to mix the fluid, that is, make the density more uniform in the vertical. In doing so light fluid is mixed down and heavy fluid up, raising the centre of gravity and increasing the gravitational potential energy. This increase in potential energy comes from the kinetic energy of the turbulence which in turn is usually derived from the kinetic energy of the mean flow. The turbulent fluid also loses some energy to heat(internal energy) through molecular viscous effects. If the rate of turbulent energy loss exceeds the rate of gain, the turbulence will die out. Indeed, if the static stability is sufficient, turbulence involving fluctuations of the vertical component will not be possible.

How can we establish a criterion for the relative importance of static stability and the tendency for instability due to the effects of the non-linear terms? As mentioned earlier, generation of turbulence requires a velocity gradient. First consider the case where v = w = 0 and u varies with z but not x or y. Then the only velocity gradient possible is $\partial u/\partial z$ and it

needs to be compared with the static stability. The possible generation of
turbulence does not depend on the sign of $\partial u/\partial z$; dynamic instability may occur
if u is either increasing or decreasing from one level to another - only a
change is required, so we consider $(\partial u/\partial z)^2$ as an indicator of the strength of
mechanical generation. A measure of the static stability is the *Brunt-Väisälä
frequency* (N) given by $N^2 = g \cdot E$ (equation 5.11). Then a measure of the
relative importance of mechanical and density effects is the dimensionless
*Richardson Number* Ri = $N^2/(\partial u/\partial z)^2$, named after the person who introduced it.
(It is sometimes called the 'gradient' Richardson Number because it is based
on gradients of mean quantities; it is possible to define a slightly different
Richardson Number based on the turbulence itself, but this extension is beyond
the scope of this book.)   If $\partial v/\partial z \neq 0$, $\partial V_H/\partial z$ would replace $\partial u/\partial z$ in the
Richardson Number.

If Ri < 0, density variations enhance the turbulence; if Ri > 0 they tend to
reduce it. If only vertical variations of $V_H$ occur and Ri becomes sufficiently
large, turbulence is not possible - the stabilizing effect of the density distribution overcomes the potential instability due to the non-linear terms.
The exact value of this 'critical' Richardson Number must be determined experimentally. This measurement is difficult to do accurately because it is
necessary to decide just when the fluid becomes barely turbulent and to account
for effects of horizontal gradients (derivatives with respect to x and y) of
velocity which are impossible to eliminate entirely. Empirically (i.e.,
experimentally) it seems that when Ri is larger than about 1/4, turbulence
cannot be generated by vertical gradients of velocity (i.e., $\partial u/\partial z$ or $\partial v/\partial z$).
Of course, if horizontal gradients of velocity are present fluctuations of
essentially horizontal velocity may develop even when Ri is much larger than
the critical value for damping vertical component fluctuations. An example is
the meandering of the Gulf Stream which is known to occur even though Ri is
probably considerably larger than its critical value.

Note that the effect of the density variation, principally in the vertical
direction, on the mean flow is indirect. It acts on the turbulence modifying
the vertical eddy viscosity (and similar coefficients for heat and salt turbulent transports). The reason is that the density variations are small (both
the fluctuations and in the mean values). Indeed there are no obvious effects
of density variation on the mean flow - there are no terms involving derivatives of mean density (or specific volume) in the equations nor are there terms
involving specific volume fluctuations. Because these fluctuations are small
we neglected the terms involving them such as $\overline{\alpha' \cdot (\partial p'/\partial x)}$ in deriving the
Reynolds equations for the mean motion, e.g., equation 7.3 for the x component.
This approximation is consistent with what is termed the *'Boussinesq approximation'*. Boussinesq said that, if the density variations are fairly small, to
a first approximation we can neglect their effect on the *mass* (i.e., inertia)
of the fluid but must retain their effect on the *weight*. That is, we must
include the buoyancy effects but can neglect the variations in horizontal
accelerations for a given force due to the mass variations with density
(which are at most 3% if we use an average over the whole ocean for $\rho$ or $\alpha$).
Thus in the horizontal momentum equations (x and y directions) we can use an
average density over the region being considered but in the z equation, which
reduces to the hydrostatic equation, we must use the actual, *in situ* values
when calculating the pressure field.

## EFFECTS OF ROTATION

In a non-rotating fluid the Reynolds and Richardson Numbers must be considered.  Once Re is sufficiently large the flow will be turbulent, at least in the
horizontal, even though a large enough Ri will restrict fluctuations of the
vertical velocity component.  Once Re is large enough for turbulence to occur
its value is not important for the mean flow directly.  The value of Re then
determines the scale size at which molecular viscous effects become important
for the turbulence itself and prevent fluctuations of smaller scales from
becoming large.

The rotation provides another possibility, that the Coriolis terms may affect
the flow.  For the ocean and atmosphere these appear to be important, often
dominant, compared with non-linear and friction terms.  The Rossby and Ekman
Numbers then determine the relative importance of non-linear effects (for the
mean flow) and frictional (usually turbulent) effects compared with the
Coriolis effects.  While the Reynolds Number based on molecular friction is
not important, the ratio of the non-linear terms (for the mean) and the
turbulent friction, measured by the Reynolds Number based on eddy viscosity,
remains an important parameter.

Having discussed the significance of the non-linear, frictional and rotational
terms in the equations of motion, we will next examine the simple approximate
equations of motion (7.9) where both the Rossby and Ekman Numbers are so
small that for the mean flow both turbulent friction (non-linear effects in
the total flow) and, except for the first example of inertial oscillations,
non-linear terms involving mean velocity components may be ignored.

# CHAPTER 8
# Currents without Friction:
# Geostrophic Flow

In this chapter we will discuss some of the characteristics of motion which we can deduce from the equations of motion when it is assumed that the $\underline{F}$ terms in equations 6.1 or 6.2 (i.e., friction, gravitation of the sun and moon, etc.) are zero and that there is a steady state, that is the velocities at any point do not change with time (i.e., $\partial u/\partial t = \partial w/\partial t = 0$). Except for the example of inertial oscillations we shall also assume that the advective acceleration terms may be neglected. These approximations for the large-scale mean circulation in the ocean's interior were justified in the previous chapter.

## HYDROSTATIC EQUILIBRIUM

As a preliminary to discussing moving fluids, let us first look at stationary ones. Let us assume that (1) $u = v = w = 0$, i.e., that the fluid is stationary, (2) $d\underline{V}/dt = 0$, i.e., the fluid remains stationary, and (3) all the $\underline{F}$ terms are zero. Then, from equations 6.2 we are left with only:

$$\alpha \cdot \frac{\partial p}{\partial x} = 0, \quad \alpha \cdot \frac{\partial p}{\partial y} = 0, \quad \alpha \cdot \frac{\partial p}{\partial z} = -g \quad . \qquad (8.1)$$

The first two mean that the isobaric (constant pressure) surfaces are horizontal, i.e., there is no pressure term, in fact no force at all in this case, to cause horizontal motion. The third can be written as

$$dp = -\rho \cdot g \cdot dz \qquad (8.2)$$

which is the *hydrostatic* (pressure) *equation* in differential form, i.e., it gives the pressure dp due to a thin layer dz of fluid of density $\rho$. If $\rho$ is constant (independent of depth) it becomes $p = -\rho \cdot g \cdot z$. This is not a very exciting result - really all that it does is confirm that the equations of motion do give a previously known answer (as shown from first principles in Appendix I), when the fluid is stationary. As we showed in Chapter 7, this equation remains an excellent approximation even for flows at typical ocean speeds.

The reason for the minus sign is because we take the origin of coordinates at the sea surface with z positive upward. Measurements up into the atmosphere are given as, for example, 'the masthead is at $+10$ m', while measurements down into the sea are given as, for example, 'the depth is 50 m' or as 'z $= -50$ m'. The pressure at this depth is (taking $\rho = 1{,}025$ kg m$^{-3}$):

$$P_{50} = -(1{,}025 \times 9.8 \times -50) = +5.02 \times 10^5 \text{ Pa} = +502 \text{ kPa} \quad .$$

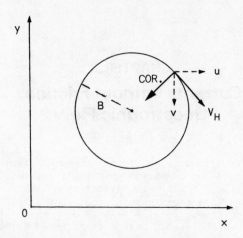

Fig. 8.1.    Relationship of Coriolis force and velocity for inertial
            motion


An increase in depth of 1 m yields an increase in pressure of about 10 kPa .


### INERTIAL FLOW

We first assume that (1) $\partial p/\partial x = \partial p/\partial y = 0$ (i.e., there is no slope of the sea
surface and all the pressure surfaces inside the fluid are also horizontal;
we shall look at the situation when these terms are not zero presently),
(2) that we can ignore the $\underline{F}$ terms as stated above, and (3) that $w = 0$
(i.e., that there is only horizontal motion).   Then the x and y equations of
motion become:

$$\frac{du}{dt} = 2\Omega \cdot \sin\phi \cdot v \qquad \text{and} \qquad \frac{dv}{dt} = -2\Omega \cdot \sin\phi \cdot u \ . \tag{8.3}$$

The equations 8.3 have solutions:

$$u = V_H \cdot \sin(2\Omega \cdot \sin\phi \cdot t)$$
$$v = V_H \cdot \cos(2\Omega \cdot \sin\phi \cdot t) \tag{8.4}$$

where $V_H^2 = u^2 + v^2$.      Now these are the equations of motion for a body in the
northern hemisphere travelling clockwise in a horizontal circle at constant
linear speed $V_H$ and angular speed $2\Omega \cdot \sin\phi$.   If the radius of the circle is
B, then  $V_H^2/B = 2\Omega \cdot \sin\phi \cdot V_H$ , i.e., the centripetal acceleration $V_H^2/B$ is
provided by the Coriolis acceleration $2\Omega \cdot \sin\phi \cdot V_H$ (Fig. 8.1).   Physically,
such motion might be generated when a wind blows steadily in one direction for
a time, causing the water to acquire a speed $V_H$, and then the wind stops and
the motion continues without friction (to a first approximation), as a consequence of its 'inertia' (properly its momentum), hence the term *'inertial
motion'*.   Flow variations of inertial period are often present in records from
current meters.   The amplitudes vary depending on the strength of generating
mechanisms and they decay due to friction when the generation stops.

Note that equations 8.3 are non-linear but do have solutions, equations 8.4, so non-linear equations can sometimes be solved explicitly.  Note also, however, that if we regard the equations as Lagrangian equations for a fluid parcel, they are linear and the terms which would be non-linear in Lagrangian terms (friction) have been assumed to be small, making solution easy.

For a speed $V_H = 0.1 \, m \, s^{-1}$ at latitude $\phi = 45°$, then $B \approx 1 \, km$.  For $V_H = 1 \, m \, s^{-1}$, then $B \approx 10 \, km$.  The period of revolution $= 2\pi/\text{angular speed} = 2\pi/2\Omega \cdot \sin\phi = $ (1 sidereal day/$\sin\phi$)/2 $= T_f/2$ because $\Omega = 2\pi/1$ sidereal day.  The quantity $T_f = $ (1 sidereal day/$\sin\phi$) is called 'one pendulum day' because it is the time required for the plane of vibration of a Foucault pendulum to rotate through $2\pi$ radians.  The value of $0.5 \, T_f$ (one-half pendulum day) is 11.97 h at the pole, 16.93 h at 45° latitude and infinity at the equator.

The direction of rotation in the inertial circle is clockwise viewed from above in the northern hemisphere and anticlockwise in the southern hemisphere.  If one thinks of observing the motion in the southern hemisphere by looking down through the earth from the northern hemisphere then the motion also appears clockwise.  However, the observer in the southern hemisphere is upside-down relative to the observer in the northern hemisphere and he calls the motion anticlockwise.  Likewise, he says that the Coriolis force acts to the left of the velocity in the southern hemisphere.  It is a matter of point of view. In the terms used by meteorologists, the motion is anticyclonic in both hemispheres.  The term *cyclonic* comes from cyclone, a storm with low pressure at its centre about which the winds are anticlockwise in the northern hemisphere and clockwise in the southern hemisphere.  An *anticyclonic* system has high pressure at its centre and winds circulate in the opposite way.  The reason for this behaviour will become clear when we discuss geostrophic flow.  Equivalent terms *contra solem* and *cum sole* are occasionally used by oceanographers in the older literature meaning, respectively, against and with the direction of motion of the sun as seen by an observer facing the equator.  These terms are related in Fig. 8.2.

### GEOPOTENTIAL

In preparation for the discussion of the geostrophic method for calculating currents we must introduce the concept of *geopotential*.  The quantity $dW = M \cdot g \cdot dz$ is the amount of work done (= potential energy gained) in raising a mass M through a vertical distance dz against the force of gravity (ignoring friction).  We then define a quantity called 'geopotential' ($\Phi$) such that the change of geopotential $d\Phi$ over the vertical distance dz is given by:

$$M \cdot d\Phi \;=\; dW \;=\; M \cdot g \cdot dz \text{ (Joules)} \quad,$$

or $\qquad d\Phi \;=\; g \cdot dz$ (Joules $kg^{-1} = m^2 \, s^{-2}$) (potential energy change/unit mass)

$$= \; - \alpha \cdot dp \text{ (from equation 8.2).}$$

Integrating from $z_1$ to $z_2$ we have:

$$\int_1^2 d\Phi \;=\; \int_1^2 g \cdot dz \;=\; - \int_1^2 \alpha \cdot dp \quad .$$

NORTHERN
HEMISPHERE

SOUTHERN
HEMISPHERE

Fig. 8.2.    Directions of rotation around low and high pressure regions in
             northern and southern hemispheres.


Now writing $\alpha = \alpha_{35,0,p} + \delta$ from Chap. 2 we get:

$$\Phi_2 - \Phi_1 = g \cdot (z_2 - z_1) = -\int_1^2 \alpha_{35,0,p} \cdot dp - \int_1^2 \delta \cdot dp \qquad (8.5)$$

$$= -\Delta\Phi_s \qquad\qquad -\Delta\Phi .$$

The quantity $(\Phi_2 - \Phi_1)$ is called the *'geopotential distance'* between the levels
$z_2$ and $z_1$ where the pressures will be $p_2$ and $p_1$. The first quantity on the
right of equation 8.5 is called the *'standard* geopotential distance' ($\Delta\Phi_s$, a
function of p only) while the second is called the 'geopotential *anomaly'*
($\Delta\Phi$, a function of S, T and p). In size, the second term is of the order of
one-thousandth of the first.

The reader is reminded that although $\Phi_2 - \Phi_1$ is called the geopotential
'distance' in oceanographic jargon, it really has the units of energy per unit
mass (J kg$^{-1}$ or m$^2$ s$^{-2}$) and for g = 9.8 m s$^{-2}$ and $\delta z = 1$ m, then $d\Phi = 9.8$ J kg$^{-1}$.
For numerical convenience, oceanographers in the past have used a unit of geo-
potential called the 'dynamic metre' such that 1 dyn m = 10.0 J kg$^{-1}$.   To
indicate that this unit is being used, it is usual to use the symbol D for
geopotential.  The geopotential distance $D_2 - D_1$ is then numerically almost
equal to $z_2 - z_1$ in metres, e.g., relative to the sea surface:

|                                                              | SI units                              | Mixed units    |
|--------------------------------------------------------------|---------------------------------------|----------------|
| at a geometrical depth in the sea                            | = + 100 m                             | + 100 m        |
| then                                                         | $z_2$ = - 100 m                       | - 100 m        |
| the pressure will be about                                   | p = + 1,005 kPa                       | + 100.5 db     |
| and the geopotential distance                                |                                       |                |
| relative to the surface                                      | $\Phi_2 - \Phi_1$ = - 980 J kg$^{-1}$, | $D_2 - D_1$ = - 98 dyn m. |

It is because of its use in the calculation of geopotential 'distance' that tables of $\alpha$ as a function of S, T and p are more common than tables of $\rho$.

## Geopotential Surfaces and Isobaric Surfaces

A surface to which the force of gravity, i.e., the plumb line, is everywhere perpendicular is called a *geopotential surface* because the value of the geopotential must be the same everywhere on the surface. The term 'level surface' is taken to mean the same thing. An example of such a surface is the smooth surface of a lake in which there are no currents and where there are no waves, or of a billiard table set up correctly. The reason for specifying 'no currents' will be explained in the next section.

An *isobaric surface* is one on which the pressure is everywhere the same. In the above stationary lake the water surface would be the isobaric surface p = 0 (atmospheric pressure being assumed constant and ignored). Isobaric surfaces for higher pressures would be deeper in the lake and would be geopotential (level) surfaces as long as the lake was still.

Isobaric surfaces must be level in the stationary state. Suppose for the moment that an isobaric surface (dashed line in Fig. 8.3a) were inclined to the level surface (full line in Fig. 8.3a). The pressure force on a particle of water A of unit mass will be $\alpha \cdot \partial p / \partial n$ as shown. ($\partial / \partial n$ means the gradient along a normal, i.e., perpendicular, to the surface and in the plane of the paper.) In addition, gravity acts on the particle. This is an unstable situation because the two forces cannot balance, as they are not exactly opposed, but must have a resultant to the left. The situation is shown in more detail for particle B where the pressure force has been resolved into:

a vertical component $\alpha \cdot (\partial p / \partial n) \cdot \cos i$ which balances g,

and a horizontal component $\alpha \cdot (\partial p / \partial n) \cdot \sin i$ which is unbalanced and would cause accelerated motion to the left, i.e., the situation is not stable.

The component to the left is

$$\alpha \cdot \frac{\partial p}{\partial n} \cdot \sin i \ = \ (\alpha \cdot \frac{\partial p}{\partial n} \cdot \cos i) \cdot \frac{\sin i}{\cos i} \ = \ g \cdot \tan i \ .$$

To stop the acceleration to the left it is necessary to apply to the right a force/unit mass, F/M, equal in size to $g \cdot \tan i$ (Fig. 8.3b). In Chapter 7 we showed that the Coriolis force was likely to be important, so one way to apply a force to the right would be to generate a Coriolis force by having the water move 'into the paper' at speed $V_1$ so that $2\Omega \cdot \sin \phi \cdot V_1$ = F/M = $g \cdot \tan i$ (Fig. 8.3c).

Fig. 8.3.    Pressure terms in relation to isobaric and to level surfaces.


## THE GEOSTROPHIC EQUATION

The Coriolis force is sometimes called the 'geostrophic' (= earth turned) force
and the equation

$$2\Omega \cdot \sin \phi \cdot V_1 \ = \ g \cdot \tan i \qquad\qquad (8.6)$$

is one version of the *geostrophic equation*.

In principle this geostrophic equation should permit us to determine the speed
$V_1$ by measuring the slope i of the isobaric surface.  In practice we cannot
determine p directly with the necessary accuracy.  Instead we have to determine
p from the hydrostatic equation $p = \int \rho \cdot g \cdot dz$ after having determined the
distribution of density $\rho$ with depth.     Even with this method we cannot
determine the angle i absolutely.  The reason is that we make our measurements
from a ship on the surface of the sea and we do not know if the sea surface is
level or not (disregarding waves).   In fact, if there are currents in the
surface waters the sea surface will *not* be level because the geostrophic
equation applies there, and motion gives rise to a Coriolis force which requires the water surface to be sloping so that the horizontal component of the
pressure gradient can act to balance the Coriolis force.  All that we can do
is determine the difference between $i_1$ at level $z_1$ and $i_2$ at level $z_2$ as
described shortly.   This difference will give us  the velocity at level $z_1$
relative to that at level $z_2$.

The slopes are small, e.g., $2\Omega \cdot \sin \phi \approx 10^{-4}$ at 45° latitude and for $V_1 = 1\,m\,s^{-1}$,
$\tan i \approx 10^{-5}$, i.e., the slope rises 1 m in 100 km, a distance typical of the
width of a strong current such as the Gulf Stream.

The geostrophic equation applies equally to the atmosphere but the meteorologist is more fortunate than the oceanographer.  He can measure air pressure directly at a number of places on the ground or at known levels in the atmosphere and then determine the horizontal pressure gradient term $[\alpha \cdot (\partial p/\partial n) \cdot \sin i]$ directly and so calculate the geostrophic wind speed.  In addition, because the speeds of currents in the ocean are small compared with wind speeds in the atmosphere, the meteorologist can ignore the water slopes and use 'mean' sea level as a reference level.

## Why Worry About the Geostrophic Equation?

The reason why the oceanographer concerns himself about using the geostrophic equation to determine currents is because direct measurement of ocean currents in sufficient quantity to be useful is technically difficult and expensive.

In shallow water a ship can anchor and hang a current meter over the side to measure the current, or can hang several meters to measure at several depths simultaneously.  However, this procedure only gives information about the currents at the one point where the ship is anchored.  Also, a ship usually does not remain stationary when anchored but moves about (i.e., it surges and swings) relative to the anchor.  Part of this motion will be added to the water motion measured by the current meter and constitutes a source of error for which it is difficult to correct.  In the deep ocean it is more difficult to anchor and the ship motion error may be much larger than the real water motion.

Many institutions are now using recording current meters which are hung in a string from a moored buoy.  A number of such buoys moored in a pattern in the ocean will provide information about the three-dimensional distribution of currents as a function of time.  However, because of the expense, the difficulties of working at sea and the complicated nature of the currents when examined in detail, it is not possible to obtain observations over as much of the ocean as we would like.

Why should we need to measure the currents over a period of time?  Why is not one measurement at each place sufficient?  Simply because real ocean currents are not steady.  They fluctuate in speed and direction and the only way to determine the mean and the variation with time is to make frequent measurements for a sufficient period of time (probably several months at least).

The geostrophic method for calculating the current requires information on the distribution of density in the ocean; it is easier to obtain this information than it is to measure currents directly.  The method suffers from several disadvantages, but when used intelligently and in parallel with other information it can be very helpful.  In fact, most of our knowledge of ocean circulation below the surface has been obtained this way.  The geostrophic method is also useful in strong currents (e.g., the Gulf Stream as we shall show near the end of this chapter) in which it is very difficult to moor recording current meters.

We should add that currents in the surface layer can be deduced from the navigation records of ships, and most of our surface layer information has been acquired from this source.  The method of using navigation records is to assume that the difference between the intended track and the one actually followed is, after correction for 'windage' (the drift of the ship due to the wind), due to the water motion.  Obviously such data are 'noisy', that is any one observation may have a large error.  By averaging over many years using

Fig. 8.4.    For the derivation of the geostrophic equation.

all the observations in a particular area (usually 5° latitude by 5° longitude) one can obtain the 'climatological' or long-term average motion.  There are undoubtedly significant variations of the actual motions from these averages; variations of several times the mean seem common according to our limited direct current observation data.  There are probably cases of smaller scale features in the flow than are resolved by such means.  For example, the pattern of flow in the equatorial Pacific deduced from observations has become more and more complicated as more detailed observations have been made. (Existing current meters do not work well near the surface, so obtaining better observations of surface currents remains a problem.)  Better navigation with the improved electronic facilities now in use will help to improve the quality of this type of data but one still does not get good coverage in time and space over most of the ocean.

The Geostrophic Method for Calculating Relative Velocities

In Fig. 8.4, A and B represent the positions where oceanographic stations have been taken so that the distribution of $\rho$ or $\alpha$ is known along each vertical $AA_1A_2$ and $BB_1B_2$.  The line AB represents the sea surface which is assumed not to be level but whose slope unfortunately is not known (and in the present state of the art is not measurable in the open sea).  $\Phi_1$ and $\Phi_2$ represent two level surfaces passing through $A_1$ and $A_2$ at station A, and $C_1$ and $C_2$ at station B.  The two isobaric surfaces $p_1$ and $p_2$ pass through $A_1$ and $A_2$ at station A, and through $B_1$ and $B_2$ at station B.  The slopes of these two isobaric surfaces are $i_1$ and $i_2$ relative to geopotential surfaces.

If the velocity component, relative to the earth, of the water (into the paper) on surface $p_1$ is $V_1$ and on $p_2$ is $V_2$, then the geostrophic equations are:

$$2\Omega \cdot \sin\phi \cdot V_1 \ = \ g \cdot \tan i_1$$

$$2\Omega \cdot \sin\phi \cdot V_2 \ = \ g \cdot \tan i_2$$

Subtracting:

$$2\Omega \cdot \sin\phi \cdot (V_1 - V_2) \ = \ g \cdot (\tan i_1 - \tan i_2)$$

i.e., $\quad 2\Omega \cdot \sin\phi \cdot (V_1 - V_2) \ = \ g \cdot \dfrac{B_1 C_1}{A_1 C_1} - \dfrac{B_2 C_2}{A_2 C_2}$

$$= \ \frac{g}{L} \cdot (B_1 B_2 - C_1 C_2) \quad \text{because } A_1 C_1 = A_2 C_2 = L$$

$$\text{and} \quad B_1 C_1 - B_2 C_2 = B_1 B_2 - C_1 C_2$$

$$= \ \frac{g}{L} \cdot (B_1 B_2 - A_1 A_2) \quad \text{because } C_1 C_2 = A_1 A_2$$

$$= \ \frac{g}{L} \cdot [(z_1 - z_3) - (z_2 - z_4)] \ . \tag{8.7}$$

Now from the hydrostatic equation:

$$g \cdot dz \qquad\qquad = \ - \qquad \alpha \cdot dp$$

$$\therefore \ \int_{B_1}^{B_2} g \cdot dz \ = \ g \cdot (z_3 - z_1) \ = \ - \int_{p_1}^{p_2} \alpha \cdot dp$$

$$= \ -\Big[ \int_{p_1}^{p_2} \alpha_{35,0,p} \cdot dp + \int_{p_1}^{p_2} \delta_B \cdot dp \Big] \ \text{from (8.5)}.$$

Note that the numerical values of the $z$'s are negative and hence $g(z_3 - z_1)$ is numerically negative as is the right-hand side of the equation.

Similarly $\qquad\qquad g \cdot (z_4 - z_2) \ = \ - \Big[ \int_{p_1}^{p_2} \alpha_{35,0,p} \cdot dp + \int_{p_1}^{p_2} \delta_A \cdot dp \Big] \ .$

Then, multiplying both equations by $-1$ to get the signs of the $z$ terms the same as in equations 8.7 and subtracting them, and noting that the two $\alpha_{35,0,p} \cdot dp$ terms are identical and therefore cancel, and dividing both sides by $L$:

$$\frac{g}{L} [(z_1 - z_3) - (z_2 - z_4)] \ = \ \frac{1}{L} \Big[ \int_{p_1}^{p_2} \delta_B \cdot dp - \int_{p_1}^{p_2} \delta_A \cdot dp \Big]$$

therefore    $(V_1 - V_2) = \dfrac{1}{L \cdot 2\Omega \cdot \sin\phi}\left[\displaystyle\int_{P_1}^{P_2}\delta_B \cdot dp - \int_{P_1}^{P_2}\delta_A \cdot dp\right]$

$\qquad\qquad\qquad\qquad\qquad\qquad\qquad\qquad\qquad\qquad\qquad$ (8.8)

$\qquad\qquad\qquad = \dfrac{1}{L \cdot 2\Omega \cdot \sin\phi}\left[\Delta\Phi_B - \Delta\Phi_A\right]$ .

(In texts using mixed units this is written:

$\qquad(V_1 - V_2) \;=\; \dfrac{10}{L \cdot 2\Omega \cdot \sin\phi} \cdot \left[\Delta D_B - \Delta D_A\right]$  where  $\Delta D \;=\; \displaystyle\int_{P_1}^{P_2}\delta \cdot dp$

with L in metres, $\delta$ in $cm^3 g^{-1}$ and p in decibars, $(V_1 - V_2)$ will be in $m\,s^{-1}$ .)

Equation 8.8 is the practical form of the geostrophic equation. The measure-ment of temperature and salinity at a series of depths at each station in principle provides the information needed to calculate the two integrals of the specific volume anomalies $\delta_A$ and $\delta_B$ (i.e., $\Delta\Phi_A$ and $\Delta\Phi_B$) while L, the distance between the stations, is obtained from navigation. In practice, the variation of specific volume anomaly with depth is never available in a methematical form suitable for direct integration and it is necessary to determine the values of the integrals (or of $\Delta D$ in the mixed system) by numerical summation (Appendix I, Integrals) as shown in the example in the next section.

If L is expressed in metres, $\delta$ in $m^3\,kg^{-1}$ , p in Pascals and $\Omega = 7.29 \times 10^5\,s^{-1}$ then $(V_1 - V_2)$ will be in metres per second. In practice it is not necessary actually to calculate the pressure from $p = -\int\rho \cdot g \cdot dz$, it is sufficient to use $p = -10^4 \cdot z$ (for z in metres, p will be in Pascals). The reason is simply that over practical distances L, say up to about 100 km, the vertical density structures are generally sufficiently similar that when the two integrals calculated using $p = -10^4 \cdot z$ are subtracted the remaining error will be neglig-ible compared to the observational errors.

The result of the calculation with equation 8.8 is a value for $(V_1 - V_2)$, the difference between the current at level $p_1$ from that at level $p_2$ *averaged* between the stations A and B. Its direction is perpendicular to the line AB and, in the northern hemisphere, would be directed 'into the paper' in Fig. 8.4 or 'out of the paper' in the southern hemisphere. With station B to the right of station A as in Fig. 8.4 a negative value of $(\Delta\Phi_B - \Delta\Phi_A)$ would indicate flow 'out of the paper' in the northern hemisphere and 'into the paper' in the south-ern hemisphere. In the figure as drawn, the slopes between the stations are constant but this need not be the case. Equation 8.8 is based on the differ-ence in distance between the isobaric surfaces *at* the two stations and there-fore on the difference in the *average* slope so it gives the average value between A and B of the difference $(V_1 - V_2)$ in the horizontal velocity component (perpendicular to AB). Note also that equation 8.8 is valid no matter what geographical direction the line AB has in the horizontal.

A better way to state the current direction on one pressure surface is to say that if this surface is sloping then, in the northern hemisphere, the current will be along the slope in such a direction that the surface is higher on the right (and *vice versa* in the southern hemisphere). Since the slopes of $p_0$, $p_1$ and $p_2$ as shown would occur if the water on the right (station B) were

TABLE 8.1a    Oceanographic Data etc. and Calculation of Geopotential Anomalies (ΔΦ) for Station A.

| Depth (m) | T°C | S‰ | $\sigma_t$ | $\Delta_{S,T}$ | $\delta_{S,p}$ | $\delta_{T,p}$ | $\delta$ | $\bar{\delta}$ | $\bar{\delta} \times \Delta p$ | $\Sigma(\bar{\delta} \times \Delta p)$ = $\Delta\Phi_A$ |
|---|---|---|---|---|---|---|---|---|---|---|
| | | Station A: 41° 55'N, 50° 09'W. | | | Units of $10^{-8}$ $m^3$ $kg^{-1}$ | | | | Units of $m^3 kg^{-1} Pa = m^2 s^{-2}$ | |
| 0 | 5.99 | 33.71 | 26.56 | 148 | 0 | 0 | 148 | | | 6.638 |
| | | | | | | | | 146 | 0.365 | |
| 25 | 6.00 | 33.78 | 26.61 | 144 | 0 | 0 | 144 | | | 6.273 |
| | | | | | | | | 135 | 0.338 | |
| 50 | 10.30 | 33.78 | 26.81 | 125 | 0 | 1 | 126 | | | 5.935 |
| | | | | | | | | 126 | 0.315 | |
| 75 | 10.30 | 34.88 | 26.83 | 123 | 0 | 2 | 125 | | | 5.620 |
| | | | | | | | | 122 | 0.305 | |
| 100 | 10.10 | 34.92 | 26.89 | 117 | 0 | 2 | 119 | | | 5.315 |
| | | | | | | | | 112 | 0.560 | |
| 150 | 10.25 | 35.17 | 27.06 | 101 | 0 | 3 | 104 | | | 4.755 |
| | | | | | | | | 99 | 0.455 | |
| 200 | 8.85 | 35.03 | 27.19 | 89 | 0 | 4 | 93 | | | 4.300 |
| | | | | | | | | 83 | 0.830 | |
| 300 | 6.85 | 34.93 | 27.41 | 68 | 0 | 5 | 73 | | | 3.470 |
| | | | | | | | | 65 | 0.650 | |
| 400 | 5.55 | 34.93 | 27.58 | 52 | 0 | 5 | 57 | | | 2.820 |
| | | | | | | | | 52 | 1.040 | |
| 600 | 4.55 | 34.95 | 27.71 | 39 | 0 | 7 | 46 | | | 1.780 |
| | | | | | | | | 45 | 0.900 | |
| 800 | 4.25 | 34.95 | 27.74 | 37 | 0 | 8 | 43 | | | 0.880 |
| | | | | | | | | 44 | 0.880 | |
| 1,000 | 3.90 | 34.94 | 27.77 | 34 | 0 | 10 | 44 | | | 0 |

lighter (less dense) than on the left (station A), the rule for direction, in the northern hemisphere, is that the current flows relative to the water below it with the 'lighter water on its right' (and *vice versa* in the southern hemisphere). (If the water is less dense at B than at A on the average, it takes a deeper column of water to produce the same pressure change ($p_0$ to $p_1$ and $p_1$ to $p_2$) as shown in Fig. 8.4.) This rule, 'light on the right', has exceptions if the vertical flow pattern is complicated but is usually true for real oceanic flows.

## An Example of the Calculation of a Geostrophic Velocity Profile

In Table 8.1 a & b, the first three columns show the depths and observed temperatures and salinities at two stations, A and B, in the region of the North Atlantic Drift (the extension of the Gulf Stream toward Europe). In successive columns are given the corresponding values at each depth for $\sigma_t$, $\Delta_{S,T}$, $\delta_{S,p}$, $\delta_{T,p}$, $\delta$ (the sum of the previous three columns), $\bar{\delta}$ the mean value of $\delta$ between each successive pair of depths (all in $m^3 kg^{-1}$), $\bar{\delta} \times \Delta p$ (in $m^3 kg^{-1} Pa = m^2 s^{-2}$) and finally $\Sigma(\bar{\delta} \times \Delta p) = \Delta\Phi$ which is the sum of the values in the previous column from each successive level to the 1,000 m level. Note

TABLE 8.1b   Oceanographic Data etc. and Calculation of Geopotential Anomalies (ΔΦ) for Station B.

| Depth (m) | T°C | S‰ | $\sigma_t$ | $\Delta_{S,T}$ | $\delta_{S,p}$ | $\delta_{T,p}$ | $\delta$ | $\bar{\delta}$ | $\bar{\delta} \times \Delta p$ | $\Sigma(\bar{\delta} \times \Delta p) = \Delta\Phi_B$ |
|---|---|---|---|---|---|---|---|---|---|---|
| | | Station B: 41° 28'N, 50° 09'W | | Units of $10^{-8}$ m³ kg⁻¹ | | | | Units of m³kg⁻¹Pa = m² s⁻² | | |
| 0 | 13.04 | 35.62 | 26.88 | 118 | 0 | 0 | 118 | | | 7.894 |
| | | | | | | | | 119 | 0.298 | |
| 25 | 13.09 | 35.63 | 26.88 | 118 | 0 | 1 | 119 | | | 7.596 |
| | | | | | | | | 119 | 0.298 | |
| 50 | 13.07 | 35.63 | 26.88 | 118 | 0 | 1 | 119 | | | 7.298 |
| | | | | | | | | 119 | 0.298 | |
| 75 | 13.05 | 35.64 | 26.89 | 117 | 0 | 2 | 119 | | | 7.000 |
| | | | | | | | | 120 | 0.300 | |
| 100 | 13.05 | 35.62 | 26.88 | 118 | 0 | 3 | 121 | | | 6.700 |
| | | | | | | | | 122 | 0.610 | |
| 150 | 13.00 | 35.61 | 26.88 | 118 | 0 | 4 | 122 | | | 6.090 |
| | | | | | | | | 122 | 0.610 | |
| 200 | 12.65 | 35.54 | 26.90 | 116 | 0 | 5 | 121 | | | 5.480 |
| | | | | | | | | 117 | 0.170 | |
| 300 | 11.30 | 35.37 | 27.03 | 104 | 0 | 8 | 112 | | | 4.310 |
| | | | | | | | | 98 | 1.980 | |
| 400 | 8.30 | 35.10 | 27.33 | 75 | 0 | 8 | 83 | | | 3.330 |
| | | | | | | | | 70 | 1.400 | |
| 600 | 5.20 | 34.93 | 27.61 | 49 | 0 | 8 | 57 | | | 1.930 |
| | | | | | | | | 52 | 1.030 | |
| 800 | 4.20 | 34.92 | 27.73 | 38 | 0 | 8 | 46 | | | 0.900 |
| | | | | | | | | 45 | 0.900 | |
| 1,000 | 4.20 | 34.97 | 27.77 | 34 | 0 | 10 | 44 | | | 0 |

that $\Delta p = -\Delta z \times 10^4$ Pa $= 10^4 \times$ depth difference in metres has been used in calculating ΔΦ.  This involves an error of up to about 1% in Δp, and hence in ΔΦ, but as the error will be almost the same for both stations, when the difference $(\Delta\Phi_B - \Delta\Phi_A)$ is taken later in the calculation it will also have an error of about 1% which is small relative to the errors due to the limited accuracy of the observations.

In Table 8.2 the values calculated for ΔΦ for stations B and A are listed and then the difference between them $(\Delta\Phi_B - \Delta\Phi_A)$.  Finally the relative speed is calculated at each depth (relative to zero speed at 1,000 m) from equation 8.8 using $L = 5 \times 10^4$ m because the stations are 27 minutes of latitude apart (= 27 n ml = 50 km), the stations being at the same longitude, and the mean value for $\sin\phi = 0.665$.

The values of the speed relative to that at 1,000 m are plotted against depth in Fig. 8.5.  The velocity component is directed to the east because relative to 1,000 m depth the isobars slope up from station A (north) to station B (south).

TABLE 8.2    Geopotential Anomalies from Table 8.1 a,b  and Calculated
Mean Relative Velocities Between Stations A and B at
Various Depths.

| Depth m | $\Delta\Phi_B$ m² s⁻² | $\Delta\Phi_A$ m² s⁻² | $(\Delta\Phi_B-\Delta\Phi_A)$ m² s⁻² | $V_{rel}$ m s⁻¹ | |
|---|---|---|---|---|---|
| 0 | 7.894 | 6.638 | 1.256 | 0.26 | Stn. A: 41° 55'N, 50° 09'W |
| 25 | 7.596 | 6.273 | 1.323 | 0.27 | Stn. B: 41° 28'N, 50° 09'W |
| 50 | 7.298 | 5.935 | 1.363 | 0.28 | Diff. =     27'      0 |
| 75 | 7.000 | 5.620 | 1.380 | 0.28 | i.e., stations are 27 n ml apart |
| 100 | 6.700 | 5.315 | 1.385 | 0.29 | = 50 km = 5 × 10⁴ m. |
| 150 | 6.090 | 4.755 | 1.335 | 0.28 | |
| 200 | 5.480 | 4.300 | 1.180 | 0.24 | sin 41° 28' = 0.662 |
| 300 | 4.310 | 3.470 | 0.840 | 0.17 | sin 41° 55' = 0.668 |
| 400 | 3.330 | 2.820 | 0.510 | 0.11 | mean sin φ  = 0.665 |
| 600 | 1.930 | 1.780 | 0.150 | 0.03 | |
| 800 | 0.900 | 0.880 | 0.020 | 0.005 | 2Ω • sin φ  = 9.70 × 10⁻⁵ |
| 1,000 | 0 | 0 | 0 | 0 | |

Fig. 8.5.   Relative velocity profile as calculated from data of Tables 8.1
and 8.2.

Note also that $(\Delta\Phi_B - \Delta\Phi_A)/g$ gives an estimate of the height difference *(dynamic topography)* of isobaric surfaces at the two stations, e.g., at the sea surface the difference is 0.13 m, that is the depths of the water from the surface to the pressure level of $10^4$ kPa (corresponding to about 1,000 m depth) differ by only 0.13 m! If the $10^4$ kPa pressure surface is also level, which it will be if the velocity there is zero, then the water surface is 0.13 m higher at station B than at station A. In the mixed units system which used dynamic metres, $\Delta D/0.98$ gave height differences in metres, so the $\Delta D$ values were numerically almost the same as the height differences. $\Delta\Phi/g$ is the difference in depth over a given pressure difference from the depth the water would have if it were standard water of S = 35‰ and T = 0°C.

## An Alternative Derivation of the Geostrophic Equation

The geostrophic equation may also be derived directly from the equations of motion (6.2) as follows:

Assume:  No acceleration, i.e., $\dfrac{du}{dt} = \dfrac{dv}{dt} = \dfrac{dw}{dt} = 0$ ,

   and   the vertical velocity,  w  =  small so that $2\Omega \cdot \cos\phi \cdot w$ may be neglected,

   and   no other forces, i.e.,  $\underline{F}$ = 0 .

Justification for these assumptions for the interior of the ocean was given in Chapter 7 .

The z equation becomes:                $0 = 2\Omega \cdot \cos\phi \cdot u - \alpha \cdot \dfrac{\partial p}{\partial z} - g$

             or:          $\delta p = -\rho \cdot \delta z (g - 2\Omega \cdot \cos\phi \cdot u)$ ,

which is the hydrostatic equation with the addition of the z component of the Coriolis acceleration. The latter, however, is small, e.g., for $\phi = 45°$, and $u = 2.5\,m\,s^{-1}$ (a high value), then the Coriolis term is about $2.6 \times 10^{-4}\,m\,s^{-2}$, which is negligible compared to $g = 9.8\,m\,s^{-2}$. Thus the hydrostatic equation (8.2) applies with all the accuracy needed even in water moving at realistic ocean speeds. This fact is fortunate as otherwise the calculation of current speeds by means of the geostrophic equation would be rendered more complicated.

Now the x and y equations become:

$\left.\begin{array}{l} x)\ 0\ =\ 2\Omega \cdot \sin\phi \cdot v - \alpha \cdot \dfrac{\partial p}{\partial x} \\[2em] y)\ 0\ =\ -2\Omega \cdot \sin\phi \cdot u - \alpha \cdot \dfrac{\partial p}{\partial y} \end{array}\right\}$  Component geostrophic equations.       (8.9)

These say that for purely horizontal motion:

or  Coriolis force + Pressure force  =  0
    Coriolis force  =  - Pressure force .

Notice that in the x and y equations we must not neglect the Coriolis term – it is small but the only other term in each equation is the horizontal pressure term which is also small but must be balanced. (One can only neglect a small term if there are other much larger terms in the equation as in the equation for the vertical component as shown above.)

Also notice that in the component equations (and see Fig. 8.6 a, b): the pressure gradient in the x direction, $\partial p/\partial x$, is associated with v, while

Fig. 8.6.    (a,b) Directional relationships of velocity components (u,v) to
             pressure and Coriolis force terms (northern hemisphere), (c)
             directional relationship of total horizontal velocity ($V_H$) to
             pressure and Coriolis force terms (northern hemisphere).

the pressure gradient in the y direction, $\partial p/\partial y$, is associated with u.    The
x and y equations can be combined into a single one:

$$2\Omega \cdot \sin\phi \cdot V_H \; = \; \alpha \cdot \frac{\partial p}{\partial n_H} \qquad\qquad (8.10)$$

  where    $V_H$   =   magnitude of the vector sum of u and v = $\sqrt{(u^2 + v^2)}$ ,

  and $\partial p/\partial n_H$   =   the horizontal pressure term perpendicular to the
           direction of $\underline{V}_H$ (see Fig. 8.6c).

One way to remember the relative direction of the pressure force and velocity
is to think of the sequence:

 (1)  the pressure gradient is initiated somehow,
 (2)  the fluid starts to move down the gradient,
 (3)  the fluid then experiences the Coriolis force to the right
    (in the northern hemisphere) and therefore swings to the
    right,
 (4)  the fluid eventually moves *along* the isobars, i.e., *along*
    the slope, not down it, with the pressure force *down* the
    slope balanced by the Coriolis force *up* the slope.

The equivalent situation in the atmosphere was shown in Fig. 8.2 to aid in
defining the terms cyclonic and anticyclonic.  It is left as a simple exercise
for the reader to verify that the circulations in this figure are consistent
with geostrophy.

Notice that an alternative procedure would be to start a fluid moving in some direction, Coriolis force would make it swing to the right (in the northern hemisphere) and pile up there (slope up to the right) so developing a pressure force to the left. Therefore the geostrophic equation simply tells us that the pressure force balances the Coriolis force - it does not tell us which came first, the pressure gradient or the motion.

Equation 8.10 is actually applicable no matter in which direction we take the pressure derivatives. If $n_H$ is taken in an arbitrary direction then $\underline{V}_H$ becomes $V_1$, the component of the geostrophic velocity perpendicular to the direction $n_H$. In the northern hemisphere, taking $n_H$ to increase to the right the flow is away from the observer if $\partial p/\partial n_H > 0$ and toward the observer if $\partial p/\partial n_H < 0$. This is another way of stating that if the isobars slope up to the right (as in Fig. 8.4) the flow is 'into the paper'.

How do we get from equations 8.9 and 8.10 to 8.8, the practical form of the geostrophic equation? The pressure derivatives in 8.9 and 8.10 are taken on surfaces of constant z which are also surfaces of constant $\Phi$. The pressure derivatives in 8.9 and 8.10 are not directly measurable, as already noted, so we must introduce the geopotential. Now using the rule from differential calculus for implicit functions

$$\left(\frac{\partial p}{\partial x}\right)_{y,z \text{ or } \Phi \text{ constant}} = -\left(\frac{\partial \Phi}{\partial x}\right)_{y,p \text{ constant}} \Big/ \left(\frac{\partial \Phi}{\partial p}\right)_{x,y \text{ constant}}$$

and remembering that $\partial\Phi/\partial p = -\alpha = -1/\rho$ , we get $\partial p/\partial x = \rho \cdot (\partial\Phi/\partial x)$ where this is the change in $\Phi$ as we go along an *isobar* in the x-direction. Likewise $\partial p/\partial y = \rho \cdot (\partial\Phi/\partial y)$ and $\partial p/\partial n_H = \rho \cdot (\partial\Phi/\partial n_H)$. These relations between p and $\Phi$ gradients can easily be obtained from first principles instead of the calculus rule. Suppose that one moves a small distance $\delta n_H$ from the point $A_1$ in Fig. 8.4. Over this distance the height on the $p_1$ isobar will change by $\delta z$, the pressure on $\Phi_1$ will increase by $\rho \cdot g \cdot \delta z$ and $\Phi$ will increase by $g \cdot \delta z$. Thus $\delta p = \rho \cdot \delta\Phi$ and dividing both sides by $\delta n_H$ and taking the limit as $\delta n_H \to 0$ gives the same relation for the derivatives as does the calculus rule. Substituting the $\Phi$ terms for the p terms in 8.9 and 8.10 gives an alternate form for the geostrophic equations.

Now these $\Phi$ gradients cannot be measured either, but differences from one level to another can be obtained from the density field. From equation 8.5 we have

$$\Phi_1 = \Phi_2 + \Delta\Phi_S + \Delta\Phi$$

and $\Delta\Phi_S$ is the same at every station, so its derivatives with respect to horizontal coordinates are always zero. Consider the x equation at levels 1 and 2 of Fig. 8.4 :

$$2\Omega \cdot \sin\Phi \cdot v_1 = \alpha\left(\frac{\partial p}{\partial x}\right)_{\text{on } \Phi_1} = \left(\frac{\partial\Phi_1}{\partial x}\right)_{\text{on } p_1} = \frac{\partial\Phi_2}{\partial x} + \frac{\partial(\Delta\Phi)}{\partial x} \text{ , and}$$

$$2\Omega \cdot \sin\phi \cdot v_2 = \frac{\partial\Phi_2}{\partial x} \text{ and the difference is}$$

$$2\Omega \cdot \sin\phi \cdot (v_1 - v_2) = \frac{\partial(\Delta\Phi)}{\partial x} \text{ .} \quad \text{Likewise}$$

$$2\Omega \cdot \sin \Phi \cdot (u_1 - u_2) = -\frac{\partial(\Delta\Phi)}{\partial y} \quad \text{and} \quad 2\Omega \cdot \sin \Phi \cdot (V_1 - V_2) = \frac{\partial(\Delta\Phi)}{\partial n_H} \quad . \quad (8.11)$$

These are differential forms of the geostrophic equation written in a way
which can be used with the kind of observations which we can make. Equation
8.8, the practical equation, appears to be a finite difference form of
equation 8.11 but in fact it is an integral form. The average along a
direction $n_H$ from 0 to L

is, by definition, $= \frac{1}{L} \cdot \int_0^L$ (quantity to be averaged) $\cdot dn_H$ . Applying this

to equation 8.11, using f for $2\Omega \cdot \sin \phi$ and an overbar to indicate an average,
gives:

$$\overline{f \cdot (V_1 - V_2)} = \frac{1}{L} \cdot \int_A^B \frac{\partial(\Delta\Phi)}{\partial n_H} \cdot dn_H = \frac{1}{L} \cdot (\Delta\Phi_B - \Delta\Phi_A) \quad . \quad (8.12)$$

The only difference between equations 8.12 and 8.8 is that averaging is not
explicitly shown in the latter and we must assume that $\overline{f \cdot (V_1 - V_2)} = \overline{f} \cdot \overline{(V_1 - V_2)}$
which it will be to a very good approximation since over the distances used in
practice f is nearly constant. In the example given in Table 8.2 with $n_H$ in
the southward direction (for which f variations with $n_H$ are a maximum), f
changed by only 1% between A and B.

## The 'Thermal Wind' Equations

These are another variation of the geostrophic equations originally derived to
show how temperature differences in the horizontal could lead to vertical
variations in the geostrophic wind velocity hence the term *thermal wind
equations*. Consider the x equation of 8.9 with f introduced and both sides
multiplied by $\rho : \rho \cdot f \cdot v = \partial p / \partial x$. Differentiation with respect to z gives:

$$\frac{\partial(\rho \cdot f \cdot v)}{\partial z} = \frac{\partial}{\partial z} \frac{\partial p}{\partial x} \quad .$$

Changing the order of differentiation, which will be correct for a variable
such as p, and using the hydrostatic equation, $\partial p / \partial z = -\rho \cdot g$, gives:

$$\frac{\partial(\rho \cdot f \cdot v)}{\partial z} = \frac{\partial}{\partial x} \cdot \frac{\partial p}{\partial z} = \frac{\partial(-\rho \cdot g)}{\partial x} = -g \cdot \frac{\partial \rho}{\partial x} \quad .$$

The same procedure can be followed for the y equation and the thermal wind
equations are:

$$\frac{\partial(\rho \cdot f \cdot v)}{\partial z} = -g \cdot \frac{\partial \rho}{\partial x}$$

$$\frac{\partial(\rho \cdot f \cdot u)}{\partial z} = g \cdot \frac{\partial \rho}{\partial y} \quad . \quad (8.13)$$

Again these show that from the density field we can only determine the vertical *variation* of velocity. The horizontal density gradients are large enough
to be observed. Because of depth uncertainties, the derivatives (based on
finite difference approximations) will not be exactly on level surfaces but
the errors are small whereas for $\partial p / \partial x$, $\partial \Phi / \partial x$, etc., they are much larger than
the actual values. In practice if one has tables for $\alpha$ rather than for $\rho$, as

is the usual case, one would use the fact that $(1/\alpha)\cdot(\partial\alpha/\partial x) = -(1/\rho)\cdot(\partial\rho/\partial x)$ and likewise for the y derivative.  In meteorology, where $\rho$ can be expressed in terms of a virtual potential temperature, gradients of this quantity may be used in place of $\rho$.  In the ocean, in the upper 1,000 m or so, as a first approximation $\partial\sigma_t/\partial x$ and $\partial\sigma_t/\partial y$ can probably be used for $\partial\rho/\partial x$ and $\partial\rho/\partial y$, respectively.  In deeper water, this is not likely to be a good approximation if temperature gradients are the dominant contribution to density gradients. (See the discussion in Chapter 5 on the use of $\sigma_t$ in the static stability equation as an approximation – the terms neglected compared with $\partial\sigma_t/\partial T$ and $\partial\sigma_t/\partial S$ are the same here as there, although they are now the coefficients of horizontal rather than vertical property gradients.)  However, in deep water, the 'thermal wind' equations are not likely to give useful results.

When we discussed the Boussinesq approximation in Chapter 7 we said that density variations could be ignored in the horizontal equations.  However, they enter the thermal wind equations because the buoyancy effects do affect the pressure field, as we also noted, and these equations were derived using the hydrostatic equation in which density variations must be included.  In the terms $\partial(\rho\cdot f\cdot v)/\partial z$ and $\partial(\rho\cdot f\cdot u)/\partial z$ the effect of density variation is small compared with the effect of vertical gradients of u and v and it would be consistent with the Boussinesq (and be a good) approximation to replace these terms with $\rho\cdot f\cdot(\partial v/\partial z)$ and $\rho\cdot f\cdot(\partial u/\partial z)$ respectively.  Here f also comes outside the derivative because it does not depend on z.

### ABSOLUTE VELOCITIES

The geostrophic calculation gives the relative velocity component $(V_1 - V_2)$. Therefore if we know the absolute value of either $V_1$ or $V_2$ we will know the absolute value of the other.  There are several possibilities:

(a) assume that there is a *level* or *depth of no motion ('reference level')* e.g., $V_2 = 0$ in deep water say, and then calculate $V_1$ for various levels above this (the classical method);

(b) when there are stations available across the full width of a strait or ocean, calculate the velocities and then apply the equation of continuity to see if the resulting flow is reasonable and complies with all facts already known about the flow;

(c) use a 'level of known motion', e.g., if surface currents are known from GEK measurements, or if the current has been measured at some depth by current meter.  (The Geomagnetic ElectroKinetograph, or GEK, invented by von Arx in 1950 is an instrument with which the motion of the surface layer is estimated relative to the earth's magnetic field as a frame of reference.)

Since surface velocities are important and can be inferred quickly from the slope of the sea surface (which is assumed to be isobaric) it is common to plot the geopotential (or dynamic) topography of the sea surface relative to some deeper surface, if a sufficient grid of station data is available.  The relative current directions will be parallel to lines of constant geopotential and relative speeds will be inversely proportional to the spacing of the lines (i.e., close spacing = steep slope = large speed, e.g., Fig. 8.7).  In the northern hemisphere, regions of high topography will be to the right when looking in the direction of the current as in Fig. 8.7.  It is also possible to plot the geopotential topography of subsurface isobaric surfaces to deduce the motion there.

SURFACE DYNAMIC TOPOGRAPHY

(Northern Hemisphere)

Fig. 8.7.  Relation between surface dynamic topography and current direction
            (northern hemisphere). (————— = lines of constant geopotential.)

Remember that these geopotential topography plots are usually based on some
assumed level of no motion, generally in the deep water, unless adequate
direct current measurements are available which is rarely the case.  If the
average upper-layer currents are much larger than the average deep currents,
which may often be the case, we may get quite good values for them even if
the deep currents are not exactly zero.

Notice that there is one known velocity region which cannot be used as a level
of no motion.  This is the sea bottom.  The reason why this depth of no motion
cannot be used is that the velocity tends to zero because of the action of
friction, a force which was deliberately assumed to be negligible when deriving the geostrophic equation.  Remember then that the geostrophic equation
does not apply in regions where friction is important.

RELATIONS BETWEEN ISOBARIC AND LEVEL SURFACES

Classically the basis for the assumption of a level of no motion was the belief that velocities are small in deep water.  Observations in recent years
with Swallow floats* have indicated that this belief may not always be correct,
and ripple marks on sand bottoms recorded in photographs in deep water suggest
that bottom currents of $0.5\,m\,s^{-1}$ or more may occur.  However, it is possible
that these indicate only local and or transient (time varying) currents and in
many regions there are indications from distributions of water properties that
speeds averaged over several days or more, or over tens or hundreds of kilo-
metres, are probably small in deep water, and the selection of a level of no

--------------------------------------------------------------
*
 These are sealed aluminum tubes which are ballasted to sink to and then float
at a predetermined level where they then travel with the water; they have a
sound source so that they can be tracked from a ship or shore station.  They
are named after their inventor, John Swallow.

Fig. 8.8.    Relations between isobaric and level surfaces.   (LNM = level of
no motion.)

motion at about 1,000 m depth may give quite good results for geostrophic
calculations.

In the Pacific, the uniformity of properties in the deep water suggests that
assuming a level of no motion at 1,000 m or so is reasonable, with very slow
motion below.   In the Atlantic, there is evidence of a level of no motion at
1,000 - 2,000 m with significant currents above *and* below this level.

A selection of relations between isobaric and constant geopotential or level
surfaces is shown in Fig. 8.8.   Fig. 8.8a is typical of the Pacific and
Fig. 8.8b of the Atlantic with the characteristics described above.   Fig. 8.8c
would indicate little motion at the surface but increasing speed into the deep
water.   This situation is unlikely in the real ocean.   Fig. 8.8d shows a
situation where all the isobaric surfaces are parallel and equally inclined

to level surfaces - the so-called 'slope current' situation.  The application
of the geostrophic calculation would yield zero relative velocity at all
depths which would be correct although the absolute velocity would not be zero.
This situation is unlikely in the ocean because density variations due to
temperature and salinity variations are likely to lead to changes in the isobaric slopes with depth.  The classical assumption is that the circulation is
mainly driven from the surface (by the wind) and the density distribution
adjusts to bring the current to zero at mid-depth.  However, it is possible
that there may be a slope current in the deep water where T and S variations
are small, plus an additional vertical variation in the upper 1,000 m or so.
If we added slopes like those of Fig. 8.8d to those of Fig. 8.8a we would
have such a situation and there would not be a level of no motion at any depth.
Observations, though they tend to be indirect, suggest that deep-water slope
currents are at least considerably smaller than near-surface currents.  While
the deep currents may have lower speeds they may transport large amounts of
water if they extend over a large depth range.  Surface currents based on
geostrophic calculations are similar to those of pilot charts (obtained independently from navigation data), which is probably the best evidence that the
average speeds of the deep waters are at least small compared with near-surface
average speeds.


## RELATIONS BETWEEN ISOBARIC AND ISOPYCNAL SURFACES

An *isobaric* surface in a fluid is one on which the hydrostatic pressure is
constant, while an *isopycnal* surface (sometimes called *isosteric*) is one on
which the density of the fluid is constant.  When the density of a fluid is a
function of pressure only [i.e., $\rho = \rho(p)$], as in fresh water of uniform
temperature, the isobaric and isopycnal surfaces are parallel to each other -
this is called a *barotropic* field of mass (Fig. 8.9a).  If the density is a
function of other parameters as well and actually varies horizontally with
them, the isobaric and isopycnal surfaces may be inclined to each other - the
*baroclinic* field (Fig. 8.9b).  This situation could occur in a freshwater lake
where the density was a function of temperature as well as pressure [$\rho = \rho(T,p)$]
or in the sea where density is a function of salinity, temperature and pressure
[$\rho = \rho(S,T,p)$].  With a barotropic field of mass the water may be stationary
but with a baroclinic field, having horizontal density gradients, such a situation is not possible.  In the ocean, the barotropic case is most common in
deep water while the baroclinic case is most common in the upper 1,000 m where
most of the faster currents occur.

It should also be noted that relative to (geopotential) level surfaces, the
isobaric and isopycnal surfaces have opposite slopes (relative to the horizontal) in the baroclinic case (assuming no barotropic flow so that the velocity
vanishes in the deep water).  Thus in the northern hemisphere, isopycnals
slope up to the left when one looks in the direction of the current.  Likewise,
isotherms generally slope up to the left in the northern hemisphere when looking in the direction of the current because temperature is generally the dominant factor in determining density in the open ocean.  The slopes will be
opposite in the southern hemisphere.

Often in oceanography barotropic flow is thought of as the flow due to a uniform tilt of the pressure surfaces like that of the deep water where the
density essentially depends only on pressure, and velocity is uniform with
depth.  Baroclinic flow is the part due to additional tilts of the pressure
surfaces caused by density variations.  For example, in Fig. 8.10 if ABC is

Fig. 8.9.    Schematic examples of barotropic and baroclinic fields of mass
             and pressure.

the vertical profile of the horizontal speed, it can be regarded as made up of
two parts, a barotropic part $V_b$ and a baroclinic part $V_c$.  The baroclinic part
may be obtained from a geostrophic calculation but the barotropic part will
not appear in this calculation, it must be obtained by some other means.  It
appears that this separation is somewhat arbitrary but it is consistent with
the general definition of a barotropic fluid that $\rho = \rho(p)$ which is true in
deep water.  However, some theoretical physical oceanographers take the surface velocity as the barotropic part and variations from this value as the
baroclinic part, so it is important to make sure which system is being used in
any particular discussion.  We shall always take the deep water, essentially
depth-independent, velocity to be the barotropic part.

Fig. 8.10.   Horizontal speed ($V_H$) as a combination of barotropic ($V_b$) and
baroclinic ($V_c$) parts.


COMMENTS ON THE GEOSTROPHIC EQUATION

The procedure of calculating the geostrophic currents from the oceanographic
data at two stations, e.g., AB (Fig. 8.11a) yields only the component (u in
this case) of current perpendicular to the line AB.  To obtain the total
current it is necessary also to make calculations for a second pair of stations,
e.g., BC, to get another component (v). These may then be added vectorially to
obtain the total current ($\underline{V}_H$) as in Fig. 8.11b.  Generally one has a grid of
stations from which groups of three can be selected to give the total current
pattern.  If one only needs the net transport through a strait, a straight
line of stations across it will be sufficient.

The geostrophic method for calculating currents suffers from several disadvantages:

(1)   It yields only relative currents and the selection of an appropriate
      level of no motion always presents a problem.  (A number of methods
      have been used and some of these are described by Sverdrup et al.
      (1946) and Defant (1960).)  However, evidence suggests that when used
      intelligently the geostrophic method gives reasonable values in deep
      water (e.g., tests by Wüst with Pillsbury's data for the Florida
      Straits, and by Knauss in the equatorial Pacific).
(2)   One is faced with a problem when the selected level of no motion reaches
      the ocean bottom as the stations get close to shore.  The (rather arti-
      ficial) procedures for dealing with this situation are discussed by
      Sverdrup et al. (1946) and Defant (1960).

Fig. 8.11.    Stations AB yield one velocity component (u), stations BC yield
              another component (v); total horizontal velocity ($\underline{V}_H$) is the
              vector sum of these.


(3)  It only yields mean values between stations which are usually many
     tens of kilometres apart.  It is impractical to place stations
     very close because:
     (a)   of the limitations on the accuracy of measurement of S, T, p
           and hence of α (differences between station values must be
           significantly larger than errors of measurement at an individ-
           ual station),
     (b)   limited navigational accuracy means that the distance (L) be-
           tween the stations may have a significant error.  Of course,
           if the ship is equipped with accurate position determining
           equipment this error will be minimised.  Satellite navigation
           can now be used to determine position to an accuracy of the
           order of 100 m or better, for a stationary ship, but drift of
           the ship while on station may introduce significant uncertainty
           in the value of L to be used in the geostrophic calculation,
     (c)   internal wave movements complicate the measurement of the
           density field by introducing periodic fluctuations for which
           it is difficult to correct.
     Actually, the fact that the geostrophic calculation only yields a mean
     value for the current over the distance between the stations introduces
     some smoothing which may not be a disadvantage if one is only interested
     in the bulk movements and does not wish to be confused by small-scale
     or short-term variations.
(4)  Friction has been ignored in deducing the geostrophic equation.  It may
     actually be significant near the bottom or where there is current shear,
     and therefore the equation does not apply in such situations.
(5)  The equation breaks down near the equator where the Coriolis force becomes
     so small that the friction forces may be important.  However, comparisons
     by Knauss with direct measurements of currents indicate that this break-
     down is only important within ±0.5° latitude (i.e., ±50 km) of the
     equator.
(6)  The calculated geostrophic current will include any long-period trans-
     ient currents and even some part of the tidal currents (although in the
     open sea the density field cannot change rapidly enough at tidal periods
     for much of the tidal current to show up in the geostrophic computation).
     It is not possible to separate the transients or tidal components from

the 'steady' ocean currents if the geostrophic current is calculated from only two stations. In principle one can repeat stations at frequent intervals and look for periodic components of the current - it is rarely practical to do so.

Despite all these disadvantages, it must be admitted that application of the geostrophic equation has provided us with much of our present knowledge of the velocities of ocean currents. It is still the only technique for obtaining information relatively quickly from a large area. Recent instrumental developments have provided means for spot observations, e.g., Swallow floats and variants give a Lagrangian picture of deep currents directly, current meters from ships held stationary near to moored buoys yield (Eulerian) current profiles and strings of current meters anchored in position are showing much promise. All these techniques require a relatively large amount of effort for relatively localized returns. The data obtained from Swallow-type floats and current meters show that there are large transient currents with many different periods which have quite complicated variations in space so that it is difficult to get a good measure of the average flow to obtain good checks of the geostrophic equation and to establish levels of known motion for geostrophic calculations over large regions.*

Some networks of anchored current meters, often with attached T, S, etc. recorders, have been located in areas of interest to permit fairly long-term studies of currents and to yield information on their variation with time. As this approach is very expensive, and probably cannot be justified economically for ocean-sized regions, perhaps the best approach will be to use numerical modelling (as described in Chapter 11) with limited measurement programmes in the ocean for adjusting the model parameters and checking the results.

## JUSTIFICATION FOR USING THE GEOSTROPHIC APPROACH TO OBTAIN THE SPEEDS OF STRONG CURRENTS

Consider the Gulf Stream as an example. It is convenient to orient the co-ordinates with the Stream so let us take the x-axis across the stream and the y-axis along it. We use scaling arguments and friction terms written in eddy viscosity form as in Chapter 7. For the width of the stream we use $L_x = 100$ km $= 10^5$ m; for the length we use $L_y = 1,000$ km $= 10^6$ m. For the along-stream, y, component of current we use $V = 1$ m s$^{-1}$ (maximum values are up to about 3 m s$^{-1}$); for the cross-stream, x, component we use $U = 0.1$ m s$^{-1}$ since the stream may spread out. Then taking the depth scale $H = 10^3$ m and using continuity, the vertical velocity is $O(U \cdot H/L_x = V \cdot H/L_y = 10^{-3}$ m s$^{-1})$. The vertical equation will reduce to the hydrostatic equation as we noted and left as an exercise in Chapter 7. Assume a steady state $(\partial u/\partial t = \partial v/\partial t = 0)$ and take maximum values for eddy viscosity $A_z = 0.1$ m$^2$ s$^{-1}$ and $A_x = A_y = 10^5$ m$^2$ s$^{-1}$ and examine the x equation:

---

*Very recently Stommel and Schott have suggested an objective method for determining the depth of no motion without the need for direct current measurements. They obtain an additional equation by assuming that there is no flow across isopycnal surfaces. The example which they show looks promising but thorough testing will be needed to see if the approach is generally applicable. If it is, then the geostrophic method would become much more useful.

$$u \cdot \frac{\partial u}{\partial x} + v \cdot \frac{\partial u}{\partial y} + w \cdot \frac{\partial u}{\partial z} = -\alpha \cdot \frac{\partial p}{\partial x} + f \cdot v - 2\Omega \cdot \cos \phi \cdot w$$

$$+ A_x \cdot \frac{\partial^2 u}{\partial x^2} + A_y \cdot \frac{\partial^2 u}{\partial y^2} + A_z \cdot \frac{\partial^2 u}{\partial z^2} \,.$$

Introducing our scales and taking $f = 10^{-4} \, s^{-1}$ (the value for $\phi = 45°$), the order of the terms is:

$$\frac{10^{-2}}{10^5} + \frac{10^{-1}}{10^6} + \frac{10^{-4}}{10^3} = ? + 10^{-4} - 10^{-7} + 10^5 \cdot \frac{10^{-1}}{10^{10}} + 10^5 \cdot \frac{10^{-1}}{10^{12}} + 10^{-1} \cdot \frac{10^{-1}}{10^6}$$

or by dividing through by $10^{-4}$, the scale for $f \cdot v$:

$$10^{-3} + 10^{-3} + 10^{-3} = ? + 1 \quad - 10^{-3} + 10^{-2} \quad + 10^{-4} \quad + 10^{-4} \,.$$

The x or cross-stream equation therefore remains geostrophic within 1%. (Remember that we have used maximum values for eddy viscosity, so 1% or so should be an upper limit for friction effects.) This equation can be used to obtain the downstream component (v) even in strong currents such as the Gulf Stream.

Now consider the y equation :

$$u \cdot \frac{\partial v}{\partial x} + v \cdot \frac{\partial v}{\partial y} + w \cdot \frac{\partial v}{\partial z} = -\alpha \cdot \frac{\partial p}{\partial y} - f \cdot u + A_x \cdot \frac{\partial^2 v}{\partial x^2} + A_y \cdot \frac{\partial^2 v}{\partial y^2} + A_z \cdot \frac{\partial^2 v}{\partial z^2}$$

i.e., $10^{-6} + 10^{-6} + 10^{-6} = - \quad ? \quad - 10^{-5} + 10^{-5} \quad + 10^{-7} \quad + 10^{-7}$ .

The non-linear terms are now about 10% of $f \cdot u$ and the largest friction term is of about the same size as $f \cdot u$. If we use a maximum value of $3 \, m\, s^{-1}$ for V, the non-linear terms become of order 30% of $f \cdot u$ and the friction terms may be up to three times $f \cdot u$. (Using a larger value for v in the x equation does not change the relative importance of the terms because v also comes into the Coriolis term.) Thus the geostrophic approximation is not good for the y equation. While we can use geostrophy to compute the downstream velocity component relative to a reference level from the density distribution, we cannot use this approximation in seeking a solution to the equations in a current as strong as the Gulf Stream. Friction, and perhaps non-linear terms, must be considered. Indeed, if friction is somewhat smaller than the maximum values (which came from estimates of friction effects in the Antarctic Circumpolar Current) non-linear terms may be comparable to or even larger than the friction terms. (In the terminology sometimes used by theoreticians, the Rossby number (Ro) and horizontal Ekman number ($E_x$ in this case) become of order one in the y momentum equation in a region such as the Gulf Stream.)

# CHAPTER 9
# Currents with Friction

A notable feature of the gross surface-layer circulation of the oceans is that it is clockwise in the northern hemisphere and anticlockwise in the southern. This fact for the North Atlantic was known to Spanish navigators in the early 1500's and was subsequently recognized for the other oceans as navigational records accumulated. In the mid-1800's this circulation was attributed to the differential solar heating between the equatorial and polar regions but no one produced any quantitative theory of the process. About 1875 Croll became convinced that this hypothesis was incorrect and suggested that the frictional stress of the wind was the direct cause, although he did not present any theory. In 1878, Zöppritz apparently demonstrated quantitatively that the transfer of momentum and energy from wind to water was much too slow a process to account for ocean currents but his demonstration was numerically in error, although he cannot be blamed. In his calculation he used the molecular co-efficient of viscosity (i.e., friction) as determined in the laboratory for laminar (smooth) flow and showed that apparently it should take months for a change of current at a depth of only a few metres to follow a change of wind at the surface. However, it was soon shown that current changes in the upper tens of metres followed wind changes in a matter of hours, not months. The reason is that in natural water bodies the flow is almost invariably turbulent and in this type of flow the turbulent or 'eddy' viscosity comes into play with the effect of increasing the vertical transfer of momentum and energy to a rate up to hundreds of thousands of times that due to molecular processes alone. This effect was not known in Zöppritz time.

Both wind driving and the effects of density changes are important for the overall circulation but the former probably dominates in the upper 1,000 m or so in most regions of the ocean. We shall discuss the wind-driven flows in this chapter and consider the differential density driving in the following chapter.

Making use of the eddy coefficient of viscosity concept, there followed a series of steps in the development of the theory of the wind-driven circulation which is now accepted, at least as a start in the right direction:

(1) about 1898 Nansen explained qualitatively why wind-driven currents flow not in the direction of the wind but at 20° to 40° to the right of it (in the northern hemisphere),

(2) in 1902 Ekman explained quantitatively for an idealized ocean how the rotation of the earth was responsible for the deflection of the current which Nansen had observed,

Fig. 9.1.    Forces on a parcel of water in the surface layer.


(3)   in 1947 Sverdrup showed how the main features of the equatorial
      surface currents could be attributed to the wind as a driving agent,
(4)   in 1948 Stommel explained the westward intensification of the wind-
      driven circulation,
(5)   in 1950 Munk combined most of the above to obtain analytic expressions
      which described quantitatively the main features of the wind-driven
      circulation in terms of the real wind field,
(6)   in recent years, numerous numerical models have been developed for
      the circulation of individual ocean areas and for the world ocean.

We will discuss these developments in the stages indicated, leaving item 6
for Chapter II.


NANSEN'S QUALITATIVE ARGUMENT

First we will present the essentials of a qualitative argument advanced by the
biologist Nansen to explain why icebergs in the Arctic drifted in a direction
to the right of the direction of the wind at the sea surface, not in the
direction of the wind itself.

In Fig. 9.1 the square represents the view from above of a cube of water in
the surface layer, while the feathered arrow indicates the wind direction.
The wind friction gives rise to a tangential force $F_t$ on the top surface of
the water tending to move it in the wind direction.    As soon as it starts to
move, the Coriolis force $F_c$ comes into action directed to the right.  In consequence the motion will be in some direction between that of $F_t$ and $F_c$.  Also
when the surface water is moving relative to that below it, there will be a
retarding force of fluid friction $F_b$ on the bottom of the cube in a direction
opposite to the motion.  The combination of $F_t$ and $F_c$ would cause the cube to
accelerate but as it does so the retarding force $F_b$ increases.  Finally, a
steady state arises in which $F_t$, $F_c$ and $F_b$ are in balance and the cube con-

Fig. 9.2.  Three forces in equilibrium on a water parcel.


tinues to move at steady speed $V_o$ in some direction between $\underline{F}_+$ and $\underline{F}_c$, i.e., to the right of the wind direction.  To determine the exact direction relative to the wind it is necessary to apply a quantitative argument from the equations of motion as Ekman did.


## THE EQUATIONS OF MOTION WITH FRICTION INCLUDED

The horizontal equations of motion become, when friction is included (and the Coriolis term involving w is omitted as noted in Chapter 6 and justified in Chapter 7):

$$\frac{du}{dt} = f \cdot v - \alpha \cdot \frac{\partial p}{\partial x} + F_x$$

$$\frac{dv}{dt} = -f \cdot u - \alpha \cdot \frac{\partial p}{\partial y} + F_y$$

(9.1)

where $F_x$ and $F_y$ stand for the components of friction per unit mass in the fluid.

If there are no accelerations (i.e., a steady state and zero, or at least neg-ligible, advective accelerations), then du/dt = dv/dt = 0 and we are left with a balance of three forces on unit mass:

$$f \cdot v \quad + \quad F_x \quad - \alpha \cdot \frac{\partial p}{\partial x} \quad = \quad 0$$

$$- f \cdot u \quad + \quad F_y \quad - \alpha \cdot \frac{\partial p}{\partial y} \quad = \quad 0$$

(9.2)

i.e.,  Coriolis + Friction + Pressure  =  0

as schematically shown in Fig. 9.2.  Remember that these are forces and must be added according to the rules for the addition of vectors.  The two equations

Fig. 9.3.  Illustrating velocity shear and absence of shear.


9.2  give the component form which would normally be used if we were doing a
numerical calculation; we can also add them graphically as in Fig. 9.2.

This situation differs from the geostrophic relation in that, with the third
force (friction) acting, the pressure and the Coriolis forces are no longer
directly opposed.  Before we can look for solutions to these equations we must
write expressions for the frictional forces $F_x$ and $F_y$.  Friction is essentially
a force which comes into being when relative motion occurs or tends to occur
between material objects.  Friction between two solid bodies is well recognized; in a fluid if two parts are in relative motion friction will also occur.
The two parts may be moving in opposite directions, or may be moving in the
same direction with one going faster than the other (Fig. 9.3).  In either
case, there is said to be 'velocity shear' in the fluid.  The amount of shear
is measured, e.g., Fig. 9.3, by: $(u_5 - u_4)/(z_5 - z_4) = \delta u/\delta z$ which tends to
$= \partial u/\partial z$ as $\delta z$ tends to 0.  Newton's Law of Friction states that in a fluid,
the friction stress $\tau$, which is the force per unit area on a plane parallel
to the flow, is given by*:

$$\tau = \mu \cdot \frac{\partial u}{\partial z} = \rho \cdot \nu \cdot \frac{\partial u}{\partial z} . \qquad (9.3)$$

The stress $\tau$ acts on the surface between the two layers which are moving at
different speeds, tending to slow down the faster and to speed up the slower.

The quantity $\mu$ is the coefficient of (molecular) dynamic viscosity, while
$\nu = \mu/\rho$ is the coefficient of kinematic molecular viscosity.  For sea water

---

*
A fluid for which the friction law of equation 9.3 holds is called 'Newtonian'.
Water, including sea water, and air behave this way but molecularly more complicated substances, such as long-chain polymers, may have a more complicated
behaviour and be non-Newtonian.

Fig. 9.4.    For derivation of the friction term in the equation of motion

at 20°C, $\mu$ has a value of about $10^{-3}\,kg\,m^{-1}\,s^{-1}$, so that $\nu$ has a value of about $10^{-6}\,m^2\,s^{-1}$.   Values vary from about 0.8 to 1.8 times these values, with temper-ature variation being mainly responsible although there is a slight salinity effect.   These are the molecular values and apply to water in smooth, laminar flow, as in a small diameter capillary tube, for Reynolds' Numbers $(Re = U \cdot L/\nu)$ of less than about 1,000 as discussed in Chapter 7.   In the ocean, where the motion is generally turbulent, the effective value of kinematic viscosity is the eddy viscosity discussed in Chapter 7 and having values of $A_x$, $A_y$ of up to $10^5\,m^2\,s^{-1}$ for horizontal shear (e.g., $\partial u/\partial y$, $\partial v/\partial x$) or of $A_z$ of up to $10^{-1}\,m^2\,s^{-1}$ for vertical shear (e.g., $\partial u/\partial z$).

The eddy friction stress $\tau = \rho \cdot A_z \cdot (\partial u/\partial z)$ expresses the force of one layer of fluid on an *area* of its neighbour above or below, but for substitution in the equation of motion we need an expression for the force on a *mass* of fluid. In Fig. 9.4 a small cube of fluid is shown with shear in the z-direction and the required force would be $(\tau_2 - \tau_1) \cdot \delta s$ in the x-direction.

As $\qquad\qquad \tau_2 \;=\; \tau_1 + \dfrac{\partial \tau}{\partial z} \cdot \delta z$

$\therefore \;\; (\tau_2 - \tau_1) \cdot \delta s \;=\; \dfrac{\partial \tau}{\partial z} \cdot (\delta s \cdot \delta z) \;=\; \dfrac{\partial \tau}{\partial z} \cdot (\delta V)$   where $\delta V$ = volume of cube.

In the limit as $\delta s$, $\delta z \to 0$, so that $\delta V \to 0$,

the force per unit volume $\;=\; \dfrac{\partial \tau}{\partial z}$ and

the force per unit mass $\;=\; \dfrac{1}{\rho} \cdot \dfrac{\partial \tau}{\partial z} \;=\; \alpha \cdot \dfrac{\partial \tau}{\partial z} \;=\; \alpha \cdot \dfrac{\partial}{\partial z} (\rho \cdot A_z \cdot \dfrac{\partial u}{\partial z}).$          (9.4)

We use $A_z$ here because we are concerned with vertical shear ($\partial u/\partial z$ or $\partial v/\partial z$). The form of equation 9.4 where $A_z$ is inside the bracket is appropriate when the eddy friction coefficient varies with depth z.   As we have very little information on the manner in which $\rho \cdot A_z$ does vary with depth, we will limit ourselves to the case where $\rho \cdot A_z$ is assumed to be constant and we can therefore rewrite equation 9.4 as:

$$\text{friction force per unit mass} \;=\; A_z \cdot \frac{\partial^2 u}{\partial z^2} \;. \tag{9.5}$$

In expression 9.4 the effect of $\rho$ variations which are very small are not important compared with variations of $A_z \cdot (\partial u / \partial z)$ and we could have taken the $\rho$ outside at this point, an approximation consistent with the Boussinesq approximation discussed in Chapter 7.

Then the horizontal equations of motion become:

$$f \cdot v + \alpha \cdot \frac{\partial \tau_x}{\partial z} \;=\; f \cdot v + A_z \cdot \frac{\partial^2 u}{\partial z^2} \;=\; \alpha \cdot \frac{\partial p}{\partial x}$$

$$\tag{9.6}$$

$$- f \cdot u + \alpha \cdot \frac{\partial \tau_y}{\partial z} \;=\; -f \cdot u + A_z \cdot \frac{\partial^2 v}{\partial z^2} \;=\; \alpha \cdot \frac{\partial p}{\partial y} \;.$$

The vertical equation reduces to the hydrostatic equation as justified in Chapter 7. The vertical velocity component, w, does not appear explicitly in the equations of motion in this form. It is obtained using the equation of continuity after first solving the equations of motion for u and v. In Chapter 7 we showed that the friction terms are small enough to be neglected in the interior of the ocean but we noted that they might not be negligible near the sea surface or bottom. For a term like $A_z \cdot (\partial^2 u / \partial z^2)$ to be significant in the equations of motion it must be comparable in size with the Coriolis term, i.e., $A_z \cdot (U/H^2) \simeq f \cdot U$. For instance, for $A_z = 10^{-1}\,m^2\,s^{-1}$, $f = 10^{-4}\,s^{-1}$ then $H^2 \simeq A_z/f \simeq 10^{-1}/10^{-4} \simeq 10^3\,m^2$ or $H \simeq 30\,m$. The friction term would still be about 10% of the Coriolis term at about $H \simeq 100\,m$, so that we should be prepared to take friction into account within this distance from the surface or bottom. (In theoretical terminology, the vertical Ekman number, $E_z$, becomes of order one near the top and bottom of the ocean.)

## EKMAN'S SOLUTION TO THE EQUATIONS OF MOTION WITH FRICTION PRESENT

A difficulty with equations 9.6 is that there are two causative forces for motion, the distribution of mass (i.e., density) which gives rise to the pressure terms, and the wind friction term. Note that we can think of the velocity as having two parts, one associated with the horizontal pressure gradient and one with vertical friction. Each part can be solved for separately and the two added together, i.e.,

$$f \cdot v \;=\; f \cdot (v_g + v_E) \;=\; \alpha \cdot \frac{\partial p}{\partial x} - A_z \cdot \frac{\partial^2 u}{\partial z^2} \tag{9.7}$$

where

$$f \cdot v_g \;=\; \alpha \cdot \frac{\partial p}{\partial z} \qquad\qquad v_g \text{ being the geostrophic velocity,}$$

and

$$f \cdot v_E \;=\; -A_z \cdot \frac{\partial^2 u}{\partial z^2} \qquad v_E \text{ being the Ekman velocity associated with vertical shear friction.}$$

This separation is possible because the equations are linear.  It provides an example of the principle of superposition, i.e., for a linear system the sum of two solutions is also a solution.  If non-linear effects become important this separation scheme does not work.

To simplify the problem, Ekman assumed the water to be homogeneous and that there was no slope at the surface, so that the pressure terms would be zero and $v_g$ therefore also zero, i.e., he solved for $v_E$ only.  He also assumed an infinite ocean to avoid the complications associated with the lateral friction at the boundaries and the diversion of the flow there.

At Nansen's suggestion, Ekman first studied the effect of the frictional stress at the sea surface due to the wind blowing over it.  Altogether he assumed:

(1)   no boundaries,
(2)   infinitely deep water (to avoid the bottom friction term),
(3)   $A_z$ constant,
(4)   a steady wind blowing for a long time
(5)   homogeneous water so that $\partial p/\partial x = \partial p/\partial y = 0$ as long as the sea surface is
       level and density depends only on pressure, i.e., a barotropic
       condition.

The reason for assumption (2) was because there was reason to believe that the wind-driven current would decrease as depth increases and therefore in very deep water the speed would become negligible.  Hence the shear would also be negligible and so the fluid friction would vanish and there would be only the friction near the surface to take into account.  The reason for assumption (3) was partly to simplify the problem and partly because so little was known about the variation of $A_z$ with z.

The equations then became:

$$f \cdot v + A_z \cdot \frac{\partial^2 u}{\partial z^2} = 0$$

Ekman's equations                    (9.8)

$$- f \cdot u + A_z \cdot \frac{\partial^2 v}{\partial z^2} = 0$$

i.e., Coriolis + Friction = 0            as in Fig. 9.5a.

If, for simplicity, we assume the wind to be blowing in the y direction (Fig. 9.5b), the solutions to Ekman's equations are:

$$u = \pm V_o \cdot \cos(\frac{\pi}{4} + \frac{\pi}{D_E} \cdot z) \cdot \exp(\frac{\pi}{D_E} \cdot z) \qquad \text{+ for northern}$$
hemisphere,
(9.9)

$$v = V_o \cdot \sin(\frac{\pi}{4} + \frac{\pi}{D_E} \cdot z) \cdot \exp(\frac{\pi}{D_E} \cdot z) \qquad \text{- for southern}$$
hemisphere.

where    $V_o = (\sqrt{2} \cdot \pi \cdot \tau_{yn})/(D_E \cdot \rho \cdot |f|)$ is the total surface current, (9.10)

$\tau_{yn}$ = magnitude of the wind stress on the sea surface (approximately proportional to the wind speed squared and acting in the direction of the wind), [$|f|$ = magnitude of f]

$D_E$ = $\pi \cdot \sqrt{2} \cdot A_z / |f|$  the *Ekman depth* or *depth of frictional influence* (discussed below).

We will interpret these solutions:

(1) at the sea surface where $z = 0$, the solutions become:

$$u = \pm V_o \cdot \cos 45°, \quad v = V_o \cdot \sin 45°$$

which means that the surface current flows at 45° to the right (left) of the wind direction in the northern (southern) hemisphere (Fig. 9.5b).

(2) below the surface, where z is no longer zero, the total current speed = $V_o \cdot \exp(\pi \cdot z/D_E)$ becomes smaller as depth increases, i.e., as z becomes more negative, while the direction changes clockwise (anticlockwise) in the northern (southern) hemisphere. The perspective drawing of Fig. 9.5c shows these two changes.

(3) the direction of flow becomes opposite to that at the surface at $z = -D_E$ where the speed has fallen to $\exp(-\pi) = 0.04$ of that at the surface. The depth $D_E$ is usually arbitrarily taken as the effective depth of the wind-driven current, the *Ekman layer*. When viewed in plan, the tips of the current vector arrows form a decreasing spiral called the 'Ekman current spiral' (Fig. 9.5d).

We made the assumption that the wind was blowing along the y direction to keep the solutions relatively simple in form for our first look at them. If the wind is blowing in some other direction the current pattern will be the same relative to the wind direction.

In order to obtain numerical relations between the surface current, $V_o$, the wind speed, W, and the depth $D_E$, we make use of two experimental observations:

Obs. 1: The wind stress magnitude $\tau_n = \rho_a \cdot C_D \cdot W^2$ where $\rho_a$ = the density of air, the drag coefficient $C_D = 1.4 \times 10^{-3}$ (non-dimensional), and W is the wind speed in m s$^{-1}$. Then $\tau_n = 1.3$ kg m$^{-3} \times 1.4 \times 10^{-3} \times W^2$ = $1.8 \times 10^{-3} W^2$ Pa. If we substitute this expression in equation 9.10 we obtain:

$$V_o = \frac{\sqrt{2} \times \pi \times 1.8 \times 10^{-3} \times W^2}{D_E \times 1025 \text{ kg m}^{-3} \times |f|} = 0.79 \times 10^{-5} \frac{W^2}{D_E \cdot |f|} \text{ m s}^{-1}. \quad (9.11)$$

Obs. 2: Field observations analysed by Ekman indicate that the surface current and the wind speed are related as:

$$\frac{V_o}{W} = \frac{0.0127}{\sqrt{\sin |\phi|}} \quad \text{outside} \pm 10° \text{ latitude from the equator.} \quad (9.12)$$

Substituting this expression in equation 9.11 (and remembering that $f = 2\Omega \cdot \sin \phi$) we get:

$$D_E = \frac{4.3 \ W}{\sqrt{\sin |\phi|}} \quad \text{metres (with W in m s}^{-1}). \quad (9.13)$$

Therefore, if we know W at latitude $\phi$ we can calculate $V_o$ and $D_E$, and the velocity at any depth below the surface. The fact that $D_E$ depends on W

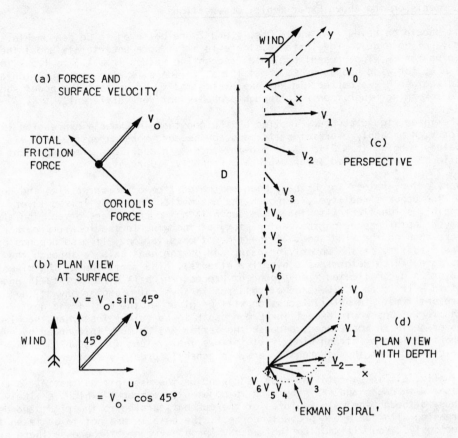

Fig. 9.5.   Wind-driven currents from Ekman analysis: (a) net frictional stress
balances Coriolis force with surface current $V_O$ perpendicular to
both, (b) wind in y-direction, surface velocity $\underline{V_O}$ and components,
(c) perspective view showing velocity decreasing and rotating
clockwise with increase in depth, (d) plan view of velocities at
equal depth intervals, and the 'Ekman spiral' (all northern
hemisphere).

suggests that the eddy viscosity $A_Z$ increases as W increases; if we know $D_E$ we
can estimate a value for $A_Z$.

Some numerical values from the above relations are given below:

|  | $\phi$ = | 10° | 45° | 80° |
|---|---|---|---|---|
|  | $V_O/W$ = | 0.030 | 0.015 | 0.013 |
| W = 10 m s$^{-1}$ : | $D_E$ = | 100 | 50 | 45 m |
| 20 m s$^{-1}$ : | $D_E$ = | 200 | 100 | 90 m |

## Comments on the Above Experimental Observations

It should be noted that while Obs. 1 and 2 are believed to be reasonable, they may not be exact. For instance, there is still some uncertainty about the value for $C_D$, the present estimates suggesting values from 1.3 to 1.5 $\times$ $10^{-3}$ $\pm 20\%$, for wind speeds up to about $15 \, m \, s^{-1}$. There is also some question as to whether $W^2$ is correct in the expression for $\tau_n$ or whether the power should be something other than 2 (although probably not very different.)

The three figure accuracy implied by the constant in Ekman's expression (9.12) for $V_0/W$ probably overstates its accuracy because more extensive data yield values ranging from 2 to 5% in mid-latitude. In addition, time-dependent effects and the mixed layer depth are probably important.

Very often, values for $D_E$ have been estimated from or compared with the depth of the upper mixed layer, although the assumption that this 'mixed layer depth' is identical with the Ekman depth is not correct very often. The mixed layer depth depends on the past history of the wind in the locality rather more than on the wind speed at the time of observation. It also depends on the stability of the underlying water and on the heat balance through the surface (which determines convective effects). The formation of the mixed layer is a complicated time-dependent process which is still not fully under-stood - it is an active area of research in physical oceanography at the present time. (A brief discussion will be given at the end of the next chapter.) One would expect the Ekman depth to be rather less than the mixed layer depth in most cases, because the latter may be much influenced by even short periods of strong winds. It follows that values for $A_z$ which are calculated from the apparent $D_E$ are, in general, probably too large.

Finally, all these detailed results depend on the assumptions that $A_z$ is con-stant with depth and that the wind is constant, neither of which is likely. Thus, although the main features of the current turning to the right and de-creasing with depth are probably correct, the details are not to be taken too seriously. You will notice that we say 'probably correct' because there are very few measured current profiles which are adequate to test the theory. It is difficult to make accurate current measurements in the open, deep ocean, the only region where the Ekman theory applies, and difficult to get suffic-iently steady wind conditions. At the same time, it must be recognized that a version of the Ekman theory also applies to the velocity structure in the atmosphere above the earth's surface and there are some observations to show that the theory applies fairly well in this case. Time dependent effects remain a problem because the wind does not often blow with a steady speed and direction for a long enough period (a few pendulum days) to produce the steady state situation. Stability may also be important.

As we mentioned in Chapter 7, in the atmosphere near the ground $A_z$ increases linearly with height at first. Higher up it is probably constant for some distance and then decreases to zero as the shear and frictional effects diminish. This layer is spoken of as the *planetary boundary* or *Ekman layer*, or sometimes the *region of frictional influence*. Assumption of a similar variation for $A_z$ in the ocean might lead to more realistic results although the presence of surface waves may make the region near the surface behave differently in the ocean than in the atmosphere. The surface waves also make measurement of currents near the surface extremely difficult and one would have to separate the Ekman flow from the geostrophic and time-varying flows (e.g., tidal), so verifying theoretical results in detail would be a non-trivial problem to say the least.

Actually, provided that the flow is reasonably uniform in the horizontal, the time-dependent problems can be solved. If u and v do not vary with horizontal position x and y, then by the continuity equation w = 0 and all the non-linear terms for the mean flow are zero. While the real flows in the ocean and atmosphere will not be exactly horizontally uniform they should be sufficiently so that the non-linear terms are negligible in most cases. Thus the equations including time variations remain linear and can be solved (by numerical methods if analytical solutions cannot be found). The problem remains of choosing a suitable behaviour for $A_z$. Solutions for the atmosphere and also for the ocean have been obtained using various dependencies of $A_z$ on z but such details are beyond the scope of this book.

## Transport and Upwelling

The wind-driven Ekman current has its maximum speed at the surface and the speed decreases with depth increase. Because the strongest currents are to the right (or left) of the wind direction, it is easy to appreciate that the net transport will be to the right (or left) of the wind direction, in fact it will be shown to be at *right angles* to the wind direction.

The basic form of the equations for horizontal motion (equations 9.6) in the absence of any pressure gradient is:

$$\rho \cdot f \cdot v + \frac{\partial \tau_x}{\partial z} = 0 \qquad\qquad \rho \cdot f \cdot v \cdot dz = - d\tau_x$$

which we can write as:     (9.14)

$$-\rho \cdot f \cdot u + \frac{\partial \tau_y}{\partial z} = 0 \qquad\qquad -\rho \cdot f \cdot u \cdot dz = - d\tau_y \ .$$

Now $\rho \cdot v \cdot dz$ is the mass flowing per second in the y direction through a vertical area of depth dz and width one metre in the x direction, and $\int_z^o \rho \cdot v \cdot dz$ will be the total mass flowing in the y direction from the level z to the surface for this strip 1 m wide, while $\int_z^o \rho \cdot u \cdot dz$ will be the total mass transport per unit width in the x direction. If we choose the lower level deep enough, then the integrals will include the whole wind-driven current. We choose a value z = $-2D_E$ where the speed will be exp($-2\pi$) = 0.002 of that at the surface, i.e., substantially zero. If we use the symbols $M_{xE}$ and $M_{yE}$ to represent the Ekman (i.e., wind-driven) transports in the x and y directions respectively, then:

$$f \cdot M_{yE} = f \cdot \int_{-2D_E}^o \rho \cdot v \cdot dz \ = \ -\int_{-2D_E}^o d\tau_x \ = \ -(\tau_x)_{Sfc} + (\tau)_{-2D_E} \qquad (9.15)$$

$$f \cdot M_{xE} = f \cdot \int_{-2D_E}^o \rho \cdot u \cdot dz \ = \ \int_{-2D_E}^o d\tau_y \ = \ (\tau_y)_{Sfc} - (\tau)_{-2D_E} \ .$$

Now $(\tau_x)_{-2D_E}$ and $(\tau_y)_{-2D_E}$ will be essentially zero because the velocity below the wind driven layer is substantially zero and therefore there can be no shear and no friction. So we have:

$$f \cdot M_{xE} = \tau_{y\eta}, \qquad f \cdot M_{yE} = -\tau_{x\eta} \qquad\qquad (9.16)$$

where we use $\eta$ to indicate surface values.

The variations of $\rho$ are small and so the $\rho$'s may be taken outside the integrals in equation 9.15 with negligible error; the value then used for $\rho$ would be a typical one, e.g., a vertical average over $2D_E$ in the region being considered.

$Q_y = \int_z^0 v \cdot dz$ is a volume transport (per unit width) which is often used as an alternative to the mass transport. Then $M_{yE} = \rho \cdot Q_{yE}$ and $M_{xE} = \rho \cdot Q_{xE}$ and alternative forms of equations 9.16 are:

$$f \cdot Q_{xE} = \alpha \cdot \tau_{y\eta}, \qquad f \cdot Q_{yE} = -\alpha \cdot \tau_{x\eta} . \qquad\qquad (9.16')$$

These results are correct even if the details of the Ekman spiral are not.

In our example, where the wind is entirely in the y-direction, $\tau_{x\eta} = 0$ and therefore $M_{yE} = 0$, but $M_{xE} > 0$ because $\tau_{y\eta} > 0$, showing that the net transport is to the right of and at right angles to the wind direction (in the northern hemisphere and vice versa in the southern hemisphere). This result remains correct for wind in any direction.

The equation of continuity then requires that there must be inflow from the left of the wind direction to replace the flow away to the right. For Ekman's infinite ocean there is no trouble in supplying this inflow in the surface layer. However, if the wind is blowing parallel to a coastline which is on the left of the wind (in the northern hemisphere) a difficulty arises. The wind causes the surface or Ekman layer to move to the right, i.e., away from the coast but because of the coast there is no supply of *surface* water on the left of the wind for replacement. What happens in nature is that, as the Ekman layer is skimmed away from the coast, water from *below* the surface comes up to replace it - this behaviour is called *upwelling* and the region near the coast is one of divergence. This phenomenon occurs at times along many regions of the eastern sides of the oceans. In the northern hemisphere the wind must blow along the coast in a southerly direction, which usually happens during the summer. In the southern hemisphere, the transport is to the left of the wind and so it must blow in a northerly direction for upwelling to occur. In general we can say what upwelling will occur when the wind blows equator-ward along an eastern boundary of the ocean in either hemisphere or poleward along a western boundary, although this latter situation is less common.

The upwelled water does not come from great depths. Studies of the properties of upwelled water indicate that it comes from depths not greater than 200-300 m. When the upwelled water has high nutrient content plankton production may be promoted and the process is therefore important biologically. Some 90% of the world's fisheries are in 2-3% of the ocean's areas, mostly in upwelling regions. However, not all subsurface waters are high in nutrient content and so upwelling does not invariably promote biological production.

If the wind blows away from the equator along the eastern boundary of the ocean, then water will be forced toward the coast and the level will rise. This process may then give rise to a surface slope and a geostrophic current. In upwelling regions also, a surface slope is usually caused, in this case down toward the coast. The induced geostrophic currents along the coast

Fig. 9.6.    (a) Convergences and divergences related to wind shear, North
             Atlantic Ocean, (b) related geostrophic flow.

generally have considerably higher speeds than the wind-induced onshore-off-
shore currents making the latter difficult to measure.  On an eastern boundary,
the downward slope toward the coast requires an equatorward flow at the
surface if the pressure gradient is to be balanced, at least mainly, by the
Coriolis force.  We say 'mainly' because near shore and/or in shallow water
friction is likely to become important and the current may not be purely geostrophic.  As the density of the water near the coast is higher than that offshore at the same level, baroclinic compensation will occur, i.e., the longshore flow will decrease with depth.  Sometimes 'overcompensation' may occur
and the offshore pressure gradient changes sign at depth, requiring a poleward undercurrent to provide a balancing (or partially balancing) Coriolis
force.

Upwelling or Downwelling Away from Boundaries

Over the real ocean, the wind is not uniform as assumed by Ekman but varies with position.  For example, if the wind remains constant in direction but varies in speed across the direction of the wind, then the Ekman transport perpendicular to the wind will vary and the upper layer waters will be forced toward or away from each other, i.e., convergences or divergences will develop. Continuity then requires that a convergence be accompanied by downward motion while a divergence be accompanied by upward motion.

For instance, in the North Atlantic the general direction of the wind is to the east at higher latitudes and to the west at lower latitudes.  (The former are called 'westerlies' because they come from the west to the observer while the latter are called 'easterlies'.)  Fig. 9.6a shows the main winds in simplified form, the lengths of the wind arrows indicating the wind speed. The Ekman transport due to the wind will be in the southward direction from the westerlies and in the northward direction for the easterlies, and the transport will be greater for greater wind speeds.  The result is that the Ekman transport to the south will increase from A to B.  To supply the in- crease, water must upwell from below the Ekman layer and there will be a zone of divergence.  From B to C the southward Ekman flow will decrease to zero and from C to D it will be in a northward direction, increasing as one goes toward D.  In consequence, the region around C will be one of convergence and water must descend below the surface.  Between D and E there will be a region of divergence and upwelling.

A wind blowing to the west along the equatorial zone will cause divergence and upwelling at the equator, because the Ekman layer transport will be to the right north of the equator and to the left south of the equator, i.e., away from the equator in both cases.

As mentioned in the preceding section there will be an additional effect.  In the region of convergence the surface level will tend to be high, while in the divergence regions it will tend to be low (Fig. 9.6b) and there will be consequent pressure gradients and geostrophic flows $u_g$ set up as shown.

Bottom Friction and Shallow Water Effects

If a current is flowing over the sea bottom, friction there will generate an Ekman spiral current pattern above the bottom but with the direction of rotation of the spiral reversed relative to the wind-driven near-surface Ekman layer.  The current pattern is shown in Fig. 9.7 for friction acting at the sea bottom, in perspective and plan views.

Assuming that $A_z$ is constant, Ekman's equations 9.8 still apply but the bound- ary conditions are different.  The tangential velocity must vanish at the bottom (i.e., u = v = 0) and must go to a constant value above the region of friction effects (the Ekman layer), assuming that the geostrophic flow above this layer is independent of z.  If, as a specific example, we take $u = u_g$ v = 0, in the geostrophic region (although again the general results of rotation relative to the direction of the geostrophic current do not depend on its direction) the solution (for the northern hemisphere) is:

$$u = u_g \cdot [1 - \exp(-\pi \cdot z/D_E) \cdot \cos(\pi \cdot z/D_E)]$$
$$v = u_g \cdot \exp(-\pi \cdot z/D_E) \cdot \sin(\pi \cdot z/D_E)$$

(9.17)

PERSPECTIVE:

Fig. 9.7.   Frictional effects on a geostrophic current near the bottom of
            the ocean (northern hemisphere).

where $z = 0$ is taken at the bottom (assumed level) in this case to make the
formulae simpler, and $D_E = \pi \sqrt{(2 \cdot A_z/|f|)}$ as before.  Equations 9.17 satisfy
9.8 (as the reader may verify by substitution).  At $z = 0$, $u = v = 0$ as required.  As z becomes large compared with $D_E/\pi$, $\exp(-\pi \cdot z/D_E)$ goes to zero
and $u = u_g$, $v = 0$ as required.  Near, but not right at, the surface $\pi \cdot z/D_E \ll 1$
and expansion of exponential, sine and cosine terms, keeping terms proportional
to $\pi \cdot z/D_E$ but neglecting higher powers, gives $u = \pi \cdot z \cdot u_g/D_E = v$.  Thus near
the bottom u and v vary linearly with z and the near-surface current direction
is 45° to the left of the geostrophic current (in the northern hemisphere and
*vice versa* in the southern hemisphere).  The current rotates from the geostrophic direction to 45° to the left of it and the speed goes to zero at the
bottom.

Using a qualitative argument similar to Nansen's for the swing of the surface
current to the right, it is easy to see why the current near the bottom swings
to the left.  Before friction begins to act we have a geostrophic current with
the Coriolis force acting to the right and the pressure gradient force to the
left.  With a barotropic case (reasonable near the ocean bottom) the pressure
gradient is independent of depth.  As the bottom is approached, friction slows
the flow; the Coriolis force (proportional to speed) decreases and the pressure gradient to the left is not completely balanced.  The flow swings to the
left until the sum of the Coriolis and friction forces can balance the pressure gradient force.

The same solution is valid (under the same assumption) for wind blowing over
the sea or land.  Since in the northern hemisphere the surface wind is at 45°

to the left of the geostrophic wind and the surface (water) current is at 45°
to the right of the surface wind, the surface surrent will be in the same
direction as the geostrophic wind (i.e., the wind above the Ekman layer). In
the southern hemisphere, the rotation directions are opposite in both cases
and the final result is the same. Again because of the simple form chosen
for $A_z$ the details should not be taken too seriously. The direction of
rotation to the left in the atmosphere is usually less than 45°, 10-20° is
more commonly observed over the ocean. This discrepancy may be due to neglect
of time-dependent and stability effects as well as to the simple form for $A_z$.
Likewise, the wind-driven surface (water) current is likely to be to the
right of the wind direction but not exactly 45°.

It is worth noting that the near-surface wind speed is still an appreciable
fraction of the geostrophic speed. At 10 m height it is 60-70% of the geostrophic speed; most of the reduction to zero occurs very close to the surface.
The Ekman layer thickness in the atmosphere is typically 10 times that in the
ocean. Thus the kinematic eddy viscosity based on this depth will be about
100 times the eddy viscosity for the ocean surface Ekman layer. This difference is a consequence of the greater speeds of flow in the atmosphere leading
to greater shears and stronger turbulent friction effects, at least as
evidenced by the value of $A_z$.

A more complicated situation might consist of a combination of a geostrophic
current with a wind-driven Ekman spiral superimposed at the surface (and
with the Ekman bottom layer if the water were shallow and the geostrophic
current extended near to the bottom). Now imagine a tidal current superimposed, the direction of which might also be rotating, and it will be appre-
ciated that the things can get quite complicated in the real ocean. It may be
very difficult to analyse into its components a current system consisting of
all three, geostrophic, wind driven and tidal, particularly if they are all
changing with time.

If you visualize the water becoming shallow and the depth decreasing to the
order of $D_E$ or less, you can see that the surface Ekman layer and the bottom
Ekman layer will close up and even overlap. In shallow water the two spirals
tend to cancel each other so that the total transport is more in the direction of the surface wind rather than at right angles to it. When the water
depth decreases to about $D_E/10$ then the transport is essentially *in* the wind
direction, the effect of the Coriolis force being swamped by the friction.

## Limitations of the Ekman Theory

The above theory is quite elegant in its way but in fact it is doubtful if
anyone has actually observed a well-developed Ekman spiral current distribution
in the sea as even Ekman in one of his last papers admitted. However, this is
not to say that the theory is incorrect - the Ekman spiral is well known and
clearly observable in the laboratory where the viscosity is molecular and constant, and there is evidence for such behaviour in the atmosphere as already
discussed. Furthermore, some of the integrated effects, such as the upwelling
consequence, are well known and common phenomena which support the Ekman
theory on broad grounds. Then why is the Ekman spiral so elusive in the sea?

The first reason is that the problem in the form solved by Ekman is very much
idealized. Commenting on his assumptions:

(1)   No boundaries - not realistic, but probably not too bad an assumption
      away from the coast, and the consequences near the coast do support
      the solution obtained.

(2)   Infinitely deep water - again not exactly true but presents only a
      small source of error in the open ocean (cf. $D_E$ values of the order of
      100-200 m compared with the average ocean depth of 4,000 m.

(3)   $A_z$ constant - probably not true but at present we do not really know
      enough about it to say whether or not this assumption leads to much
      error.  It probably does not, because Rossby *et al.* have solved the
      equations with $A_z = f(z)$ in likely ways and found only detailed differ-
      ences from Ekman's solution, e.g., the angle between the wind and the
      current at the surface became slightly smaller and a function of
      latitude and wind speed.

(4)   Steady state solution and steady wind - probably a real source of
      difficulty, since neither wind nor sea is really steady (except approxi-
      mately in the trade wind zones).  Furthermore, there are other sources
      of motion in the sea (thermohaline, tidal, internal waves) and a
      current meter placed in the sea cannot distinguish one from another.
      It records the sum and the oceanographer has to try to sort them out.
      To do so it is necessary to have long series of measurements (say
      hourly or even more frequently, for months); these we lack in suffic-
      ient detail to test Ekman's theory adequately.  Added to this deficiency
      are the practical difficulties of measuring currents in deep water, the
      only region where it is reasonable to seek the Ekman spiral.

(5)   Homogeneous water - distinctly unreal and one assumption that should be
      criticised although as noted the wind friction part of the flow can be
      calculated separately.  Sverdrup was probably the first to try to do
      something to correct this fault, as will be described in the next
      section.

Despite its idealized nature this theory of wind-driven currents, stimulated
by Nansen and worked out by Ekman, opened the way to the understanding of the
mechanism giving rise to the upper-layer currents.  The key to Ekman's success
here was the use of the large eddy coefficient of viscosity rather than the
much smaller molecular one which rendered Zöppritz' earlier attempt sterile.

## SVERDRUP'S SOLUTION FOR THE WIND-DRIVEN CIRCULATION

The equations of motion assuming negligible accelerations and friction from
horizontal gradients of velocity are:

$$\alpha \cdot \frac{\partial p}{\partial x} = f \cdot v \quad + \alpha \cdot \frac{\partial \tau_x}{\partial z}$$

$$\alpha \cdot \frac{\partial p}{\partial y} = -f \cdot u \quad + \alpha \cdot \frac{\partial \tau_y}{\partial z} \ ,$$

(9.6')

i.e.,   Pressure = Coriolis + Friction  (forces) .

Ekman simply ignored the pressure gradient terms on the left side, assuming an
unrealistic homogeneous ocean with level isobars.  Here we ignore horizontal
friction terms which would be important in currents such as the Gulf Stream,
so the solutions are not valid there.  However, we have added wind driving

and can examine its possible effects away from coastal boundaries.

Essentially what Sverdrup did was to retain the pressure terms but abandon
any attempt to determine the details of the velocities u and v as a function
of z.  He was satisfied to determine the total transport in the x and y
directions in the whole layer affected by the wind ($M_x$ and $M_y$ when expressed
as mass transport).  He integrated the equations from z = - h (assumed to be
above the ocean bottom) where the wind-driven motion had become zero.  Such
motion would include not only the Ekman flow but any geostrophic flows caused
by divergence of the Ekman flow, so h >> $D_E$.  In the first stage of integration the equations take the form:

$$\int_{-h}^{0} \frac{\partial p}{\partial x} \cdot dz = \int_{-h}^{0} \rho \cdot f \cdot v \cdot dz + \tau_{x\eta} = f \cdot M_y + \tau_{x\eta}$$

$$\int_{-h}^{0} \frac{\partial p}{\partial y} \cdot dz = -\int_{-h}^{0} \rho \cdot f \cdot u \cdot dz + \tau_{y\eta} = -f \cdot M_x + \tau_{y\eta} .$$

(9.18)

Here, $\tau_{x\eta}$ and $\tau_{y\eta}$ represent the wind friction stress at the sea surface, all
that remains of the friction terms in the previous pair of equations 9.6'.
The reason is that when the derivative of a continuous function is integrated
between limits, the values *at* the two limits determine the value of the integral.  In this case the value of the friction stress in the water is equal
to the wind stress at the surface, taken to be at z = 0 (the τ values) and is
zero at z = - h because it was selected to be where the motion had become zero,
and with no motion in a fluid there is no friction.  To simplify the notation
we shall omit the η subscript in the rest of this section and use $\tau_x$ and $\tau_y$
for the surface stress components.

We shall carry out a more general derivation which includes Sverdrup's simpler
one as a special case presently, so we shall use some results without proof
here.  If we differentiate the first of equations 9.18 with respect to y and
the second with respect to x, then the differentiation of the pressure terms
can be taken inside the integrals since the limits are constants.  (The
surface is not exactly level but this variation does not matter, as we will
show later, and we ignore it.)  The two pressure terms are then the same
except for the order of differentiation which can be interchanged for a variable such as p, so the pressure terms are the same.  Then subtracting the two
equations, noting that the term $M_x \cdot (\partial f/\partial x)$ which appears is zero (because
the Coriolis term, f, does not change in the x-direction (east-west) and that
$(\partial M_y/\partial y) + (\partial M_x/\partial x) = 0$ by continuity, the resulting equation is:

$$M_y \cdot \frac{\partial f}{\partial y} = \frac{\partial \tau_y}{\partial x} - \frac{\partial \tau_x}{\partial y}$$

(9.19)

and, together with the equation of continuity for mass transport

$$\frac{\partial M_y}{\partial y} + \frac{\partial M_x}{\partial x} = 0 ,$$

(9.20)

these form a pair of equations describing the mass transports $M_x$ and $M_y$.

Sverdrup's procedure assumes either that there is zero velocity in the deep water or that the bottom is level and that the friction there is small com-pared with that at the surface.  If there is a barotropic current and a variable bottom depth where the water is moving there will be additional terms which we shall discuss later.  The interesting feature of equation 9.19 is that it is not the components themselves, $\tau_x$ and $\tau_y$, of the wind surface stress $\underline{\tau}_n$ which appear but their horizontal *gradients* $\partial\tau_x/\partial y$ and $\partial\tau_y/\partial x$.  In equation 9.19 the combination $(\partial\tau_y/\partial x - \partial\tau_x/\partial y)$ is the vertical component $(\text{curl}_z \underline{\tau}_n)$ of the curl of the wind stress $(\nabla \times \underline{\tau}_n)$, the only component which is non-zero for a horizontal wind.  The symbol $\beta$ is often used for $\partial f/\partial y$ and with these changes of notation, equation 9.19 becomes:

$$\beta \cdot M_y = \text{curl}_z \underline{\tau}_n \tag{9.21}$$

which is called the *Sverdrup equation*.

At some places $\text{curl}_z \underline{\tau}_n$ will vanish (equal zero) and there will be no north-south transport (although there may be flows which when added up cancel). Lines along which $\text{curl}_z \underline{\tau}_n = 0$ provide natural boundaries which separate the circulation into 'gyres'.

The quantities $M_x$ and $M_y$ are the total mass transports in the wind-influenced layer, defined as $M_x = \int_{-h}^{0} \rho \cdot u \cdot dz$ and $M_y = \int_{-h}^{0} \rho \cdot v \cdot dz$. We write

$$M_x = M_{xE} + M_{xg}$$

where the first term on the right is the Ekman wind-driven transport while the second is the geostrophic transport, and similarly for $M_{yE}$ and $M_{yg}$ (just as we separated v into $v_E$ and $v_g$).

Then equations 9.18 become:

$$f \cdot M_{yE} = f \cdot \int_{-h}^{0} \rho \cdot v_E \cdot dz = -\tau_x$$

$$\tag{9.22}$$

$$f \cdot M_{yg} = f \cdot \int_{-h}^{0} \rho \cdot v_g \cdot dz = \int_{-h}^{0} \frac{\partial p}{\partial x} \cdot dz \; ,$$

and similarly for $M_x$.

## Orders of Magnitude of the Terms

It is useful to look at the magnitudes of some of the terms.  We use a position in the North Atlantic at about 35° N with a wind of $7-8\,\text{m s}^{-1}$ (about 15 knots) from the west.  Then $\tau_x \approx 10^{-1}\,\text{N m}^{-2}$ (or Pa) and $\tau_y = 0$,

$$\text{curl}_z \underline{\tau}_n \approx -\frac{\partial\tau_x}{\partial y} \approx -\frac{10^{-1}\,\text{N m}^{-2}}{1000\;\text{km}} \approx -10^{-7}\,\text{N m}^{-3}$$

$$f \approx 10^{-4}\,\text{s}^{-1}, \quad \beta \approx 2 \times 10^{-11}\,\text{m}^{-1}\,\text{s}^{-1}.$$

From these values we get, using equation 9.16:

Fig. 9.8.    For derivation of the Sverdrup transport on the east side of
an ocean.

$$M_{yE} \; = \; -\frac{\tau_x}{f} \; = \; -10^3 \; \text{kg m}^{-1} \, \text{s}^{-1}$$

(where the − sign indicates flow to the south) and, using equation 9.21:

$$M_y \; = \; M_{yE} + M_{yg} \; = \; \frac{\text{curl}_z \, \tau_n}{\beta} \; = \; -\frac{10^{-7}}{2 \times 10^{-11}} \; = \; -5 \times 10^3 \, \text{kg m}^{-1} \, \text{s}^{-1} .$$

We see that $M_{yg}$ = $-4 \times 10^3 \, \text{kg m}^{-1} \, \text{s}^{-1}$ which is caused by the north-south variation of the wind and consequent convergence of $M_{yE}$ and is considerably larger
than $M_{yE}$ as is often the case.

The value for $M_y$ above is for only a 1 m wide strip, so that for the width of
an ocean of 5,000 km = $5 \times 10^6$ m, the southward flow would be $25 \times 10^9 \, \text{kg s}^{-1}$
$\approx 25 \times 10^6 \, \text{m}^3 \, \text{s}^{-1}$ in volume = 25 Sverdrups.  (Here, as is common practice in
oceanography, we use 1,000 $\text{kg m}^{-3}$ for $\rho$ when converting from mass to volume
transport since the error is negligible compared with the uncertainty in the
estimates of transport.)

Application of the Sverdrup Equations

Sverdrup applied these equations to the trade-wind zones in lower latitudes
where $\tau_y$ and $\partial \tau_y / \partial x$ can be assumed to be negligible (i.e., much smaller than
the terms retained), and $\tau_x$ variations with x are averaged out.  Substituting
f = $2\Omega \cdot \sin \phi$ and noting that y = $R \cdot d\phi$ where R = radius of the earth, then
$\beta$ = df/dy = $d(2\Omega \cdot \sin \phi)/R \cdot d\phi$ = $2\Omega \cdot \cos \phi/R$ and equation 9.19 gives $M_y$ while
using 9.20 gives

$$\frac{\partial M_x}{\partial x} \; = \; -\frac{\partial M_y}{\partial y} \; = \; \frac{1}{2\Omega \cdot \cos \phi} \left( R \cdot \frac{\partial^2 \bar{\tau}_x}{\partial y^2} + \frac{\partial \bar{\tau}_x}{\partial y} \cdot \tan \phi \right)$$

which can be integrated from x = 0, the coast, where $M_x$ = 0 .   Finally we have:

Fig. 9.9.   Smoothed representation of the east-west components of wind speed
            and stress term, and related currents at low latitudes, eastern
            Pacific Ocean.

$$M_x = \frac{x}{2\Omega \cdot \cos\phi} \left( \frac{\partial\bar{\tau}_x}{\partial y} \cdot \tan\phi + \frac{\partial^2\bar{\tau}_x}{\partial y^2} \cdot R \right) ,$$

$$M_y = \frac{-R}{2\Omega \cdot \cos\phi} \cdot \frac{\partial\bar{\tau}_x}{\partial y} .$$

(9.23)

Here, x is the distance from a north-south coastline at the east side of the
ocean westward to a point P in the ocean as in Fig. 9.8, so that the numerical
value entered in the expression will be negative. The bars over the stress
terms indicate that mean values are taken over the distance x and the values
for $M_x$ and $M_y$ are for the point P.

Comparing the Ekman and the Sverdrup solutions, Sverdrup lost the details of
the current velocities with depth but gained the possibility of having a
coastal boundary at one side of the ocean, a step toward a more realistic
situation than Ekman's horizontally infinite, i.e., boundaryless, ocean.
Sverdrup's solution also is no longer bound by the homogeneous ocean assumption
and the solutions therefore have this additional feature of the real oceans.

Referring to the expression for $M_x$ in equation 9.23 above, it turns out in
practice that in the trade wind and equatorial zones, the important term of
the two on the right is $\partial^2\bar{\tau}_x/\partial y^2$. Figure 9.9 shows for the eastern Pacific
the character of the mean x-component of the wind as the full line while the
corresponding character of $\partial^2\bar{\tau}_x/\partial y^2$ is shown by the dashed line. By 'character'
we mean schematically - the actual wind variation with latitude is not as
regular as shown.

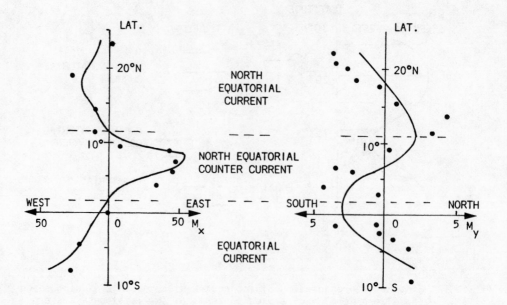

Fig. 9.10.    Mass transport components in the eastern Pacific calculated from
              the mean wind stress (lines) compared to those from geostrophic
              calculations from oceanographic data (dots). $M_x$ and $M_y$ in tonnes
              per second through a vertical area 1 m wide and 1,000 m deep
              (approximately equivalent to 0.1 Sv per degree latitude). (From
              R.O. Reid, *J. Mar. Res.*, 7, 90 & 91, 1948.)


It will be seen that:

    (a)   north of about 15°N and south of about 2°N, the value of
          $\partial^2 \bar{\tau}_x / \partial y^2$ is +  and  x is - ,   $\therefore$  $M_x$ is - ,
          i.e., flow is to the WEST (North Equatorial Current and
          Equatorial Current).

    (b)   between 15°N and 2°N (doldrums), the value of
          $\partial^2 \bar{\tau}_x / \partial y^2$ is -  and  x is - ,   $\therefore$  $M_x$ is + ,
          i.e., flow is to the EAST (North Equatorial Counter-Current).

The figure shows qualitatively how Sverdrup's solution explains the existence
of the equatorial current system consisting of two westward flowing currents
(N.E.C. and E.C.) with an eastward flowing current (N.E.C.C.) between them.
It will be noted that this system is not symmetrical about the equator but is
displaced to the north of it, because the trade wind system is displaced this
way.  (It should be mentioned that Sverdrup referred to the three currents as
the North Equatorial Current, the Equatorial Counter Current and the South
Equatorial Current, in order from north to south.  However, the equatorial
current system is now known to be more complicated than was recognized by
Sverdrup and we have used the present names for the three currents.)

Fig. 9.11.    Streamlines of mass transport in the eastern Pacific from the
mean wind stress (from R.O. Reid, *J. Mar. Res.*, 7, 95, 1948).

Sverdrup went on to test the solutions quantitatively by:

(a)  calculating $\tau_x$ from the known mean winds and then calculating the curl,
     etc., and thence values for $M_x$ and $M_y$ at selected positions defined by
     $x$ (distance from the eastern boundary),

(b)  determining $M_{xg}$ and $M_{yg}$ independently by the geostrophic method from
     oceanographic data and adding the Ekman transport to get the total
     transport,

(c)  comparing these two independent calculations.

The result of this calculation, as revised by Reid, is shown in Fig. 9.10
where the values for $M_x$ are shown on the left and for $M_y$ on the right. Because
$M_y$ is smaller, the relative errors in obtaining it from the density field are
greater which may be at least part of the reason for the poorer agreement in
the two methods for $M_y$ than for $M_x$. Note that $\partial^2 \bar{\tau}_x / \partial y^2$, which is proportional
to $M_x$, has a more complicated structure when the actual winds are used than
that in the schematic picture shown in Fig. 9.9.

Note that $M_x \approx 10 \, M_y$ which is fairly typical, particularly for equatorial
regions. The reason lies in the difference between the east-west and north-
south length scales of the gyre systems. The east-west scale ($L_x$) is determined by continental barriers, the north-south ($L_y$) by the lines of $\text{curl}_z \, \tau_n = 0$.
Typically there is approximately a 10 to 1 ratio of lengths $L_x/L_y$. Because by
continuity

$$\frac{\partial M_x}{\partial x} + \frac{\partial M_y}{\partial y} = 0, \quad \text{then} \quad \frac{M_x}{M_y} \approx \frac{L_x}{L_y} \approx \frac{10}{1} \, .$$

Looking at it in another way, water which goes north or south in a gyre must then go east or west to close the gyre. Thus the total transport north-south equals that east-west, i.e., $M_y \cdot L_x = M_x \cdot L_y$. This gives the same result using the idea of continuity of volume in an integral rather than a differential sense.

Figure 9.11 shows an analytic solution by R.O. Reid calculated from a simplified form for the wind stress but for the real coastline in the eastern equatorial Pacific. Presentation of flow patterns in this way is discussed in Appendix I. The stream function $\psi$ will be defined presently. For the moment, take it to be a horizontally integrated mass transport. The water flows in the direction indicated by the arrows on the lines; between any two lines the total transport is $5 \times 10^6$ tonnes per second (or about 5 Sv).

The calculation (a) above in Sverdrup's test was from the mean wind over a long period while the calculation (b) was for specific cruise data. However, the equatorial current system seems to be a permanent feature and the trade wind pattern is also a regular feature. Therefore the correspondence between the known current patterns and that obtained from the wind stress *via* Sverdrup's theory is taken as strong support for the theory, which is now accepted as providing the basic explanation for the equatorial current systems in each of the oceans, and also for its application to other parts of the ocean.

At the same time one must recognize the limitations of the Sverdrup theory as he applied it:

(1)  It is limited in application to the neighbourhood of the east coast of the ocean, because the x in the expression for $M_x$ (equation 9.23) would appear to make $M_x$ increase in direct proportion to the distance to the west. $M_x$ does increase somewhat to the west but not as fast as the expression would suggest. Probably the reason for this discrepancy is that lateral friction (between the currents) has been ignored. It will increase as the currents increase and therefore in the real ocean $M_x$ does not go on increasing to the west as rapidly as the Sverdrup relation suggests. The stress terms $\tau_x$ and $\partial^2 \tau_x / \partial y^2$ no doubt also have some x variation which was not included.

(2)  The differential equations allow only one boundary condition to be satisfied; in the solution given it is that there shall be no flow through the coast. To be able to apply more boundary conditions (e.g., no slip on the eastern boundary and perhaps conditions at a western boundary) it is necessary to go to more complicated equations, as will be described later.

(3)  The solutions give the integrated mass transport but no details of the velocity distribution with depth.

## THE GENERAL FORM OF THE SVERDRUP EQUATION

As we have already seen when using mass transports, we need an equation of continuity for them. Consider a column of fluid with sides $\delta x$ and $\delta y$ extending from the sea surface to $z = -h$. To be completely general we denote the value of z at the surface by $\eta$ because we know that the surface may have small variations from the level surface $z = 0$ which we may take to be the average level of the sea surface over the region being examined. Mass transports in the x and y directions respectively are:

$$M_x = \int_{-h}^{n} \rho \cdot u \cdot dz \quad \text{and} \quad M_y = \int_{-h}^{n} \rho \cdot v \cdot dz \ .$$

The mass flow into the column in the x-direction is $M_x \cdot \delta y$ and the flow out is $[M_x + (\partial M_x/\partial x) \cdot \delta x] \cdot \delta y$ . (This calculation is just like that used in the derivation of the equation of continuity in Chapter 4, except that now we have a column of height $(h + \eta)$ instead of height $\delta z$.) The net flow out in the x-direction is $(\partial M_x/\partial x) \cdot \delta x \cdot \delta y$ . Similarly in the y-direction it is $(\partial M_y/\partial y) \cdot \delta x \cdot \delta y$ . At the bottom of the column $(z = -h)$ there may be flow out if the vertical velocity is not zero; it will be $-\rho \cdot w_{-h} \cdot \delta x \cdot \delta y$ (where the minus sign is required because we want flow *out* while w positive is *up*). At the top there may be an effective outflow velocity $w_n$ if there is a net difference between evaporation and precipitation. We shall neglect this effect but it could be included as a driving term. If the value of $\eta$ is changing with time then the equivalent mass flow out is $\rho \cdot (\partial \eta/\partial t) \cdot \delta x \cdot \delta y$ . As we are considering steady state cases we shall take it to be zero. (It would be important in time-dependent calculations for such phenomena as tides and storm surges.) Because mass must be conserved the net flow of mass out must be zero, i.e.:

$$\left( \frac{\partial M_x}{\partial x} + \frac{\partial M_y}{\partial y} - \rho \cdot w_{-h} \right) \cdot \delta x \cdot \delta y = 0$$

and dividing through by $\delta x \cdot \delta y$ gives:

$$\frac{\partial M_x}{\partial x} + \frac{\partial M_y}{\partial y} - \rho \cdot w_{-h} = 0$$

(9.24)

or

$$\nabla_H \cdot \underline{M} \quad - \rho \cdot w_{-h} = 0$$

where $\nabla_H \cdot = [\underline{i} \cdot (\partial/\partial x) + \underline{j} \cdot (\partial . \partial y)]$ is the horizontal divergence operator and $\underline{M}$ is the vector mass transport.

If the velocity is zero at $z = -h$ and deeper, then the last term of equation 9.24 is zero and we get the form used by Sverdrup. If $h = -z_B$ is the total depth (which in general will be variable) then the last term also vanishes because there can be no flow through the bottom.

Sometimes it is convenient to take the point of view that the vertical velocity associated with the divergence of one type of transport provides a driving force for the convergence of another type of transport. For example, consider the Ekman flow, then $(\partial M_{xE}/\partial x) + (\partial M_{yE}/\partial y) - \rho \cdot w_E = 0$ . Here, $w_E$ is the vertical velocity at the bottom of the Ekman layer associated with convergence or divergence of the Ekman transport. If $w_E \neq 0$, it requires a corresponding divergence or convergence of the flow below. Denoting this flow by $\underline{M_g}$ (g for geostrophic) and assuming for the present that the flow vanishes below $z = -h$, so that $w_{-h} = 0$, then $\nabla_H \cdot \underline{M} = \nabla_H \cdot (\underline{M_E} + \underline{M_g}) = 0$ . Thus $\nabla_H \cdot \underline{M_g} = -\nabla_H \cdot \underline{M_E} = -\rho \cdot w_E$ . The process is sometimes spoken of as *Ekman pumping*. If the flow does not go to zero before approaching the bottom, there may be a bottom Ekman layer. Then $\nabla_H \cdot \underline{M_{EB}} + \rho \cdot w_{EB} = 0$ where $\rho \cdot w_{EB}$ is added now since it is flow up through the top and $\nabla_H \cdot \underline{M_g} = -\rho(w_E - w_{EB})$ . We expect $w_{EB}$ to be rather small compared with $w_E$ as a general rule. Note that $w_E$ is a good approximation to the total vertical velocity at the base of the layer. There may be divergence of the part of $M_g$ occurring in the relatively thin Ekman layer. The total divergence $\nabla_H \cdot M$ is only $-\rho \cdot w_E$ and, since the Ekman layer is only a small fraction of the region of geostrophic flow, the divergence of $M_g$ in the Ekman layer will be a small part of the total and may be neglected.

If we use volume transports, $Q_x = \int_{-h}^{\eta} u \cdot dz$ and $Q_y = \int_{-h}^{\eta} v \cdot dz$, and assume incompressibility, the net volume flow out must vanish. By the same sort of derivation and neglecting the possible flows at the surface, we have:

$$\frac{\partial Q_x}{\partial x} + \frac{\partial Q_y}{\partial y} - w_{-h} = 0 . \tag{9.25}$$

Let us now integrate the equations of motion 9.6 vertically from the bottom $z = z_B$ to the surface $z = \eta$. In the equations 9.6 accelerations and friction from velocity variations in the horizontal have been assumed small; this should be a good approximation except in strong currents (usually near the western boundary) such as the Gulf Stream. We obtain:

$$\int_{z_B}^{\eta} \frac{\partial p}{\partial x} \cdot dz = f \cdot M_y + \tau_{x\eta} - \tau_{xB}$$

$$\int_{z_B}^{\eta} \frac{\partial p}{\partial y} \cdot dz = -f \cdot M_x + \tau_{y\eta} - \tau_{yB} \tag{9.26}$$

where we have retained the possibility of friction occurring at the bottom although we expect it not to be important. The symbol $\eta$ in the subscript for a stress component indicates a surface value while B denotes a value at the bottom.

We cannot directly follow Sverdrup's procedure of cross-differentiating the pressure terms because the limits of integration are not independent of x and y now. To get around this problem we must write the pressure terms in a somewhat different form. Here we follow Fofonoff's article in *The Sea*, Vol. 1 (M.N. Hill, Ed., 1962) and do not derive all the results in detail; the interested reader should consult the article which also contains other valuable information. We define a new function $E_p = \int_{z_B}^{\eta} p \cdot dz$. It can be shown (by integrating by parts taking p = 0 at the surface and using the hydrostatic equation) that:

$$E_p = \int_{z_B}^{\eta} \rho \cdot g \cdot (z - z_B) \cdot dz = \text{work done to pile up the water above the bottom,}$$

i.e., it is a measure of the potential energy. Now

$$E_p = \frac{1}{g} \cdot \int_{z_B}^{\eta} (p \cdot \alpha) \cdot \rho \cdot g \cdot dz \quad \text{is the original definition rewritten.}$$

Using the hydrostatic equation $dp = -\rho \cdot g \cdot dz$ and p = 0 at $z = \eta$, $p = p_B$ at $z = z_B$ we have:

$$E_p = \frac{1}{g} \cdot \int_0^{p_B} p \cdot \alpha \cdot dp = \frac{1}{g} \cdot \int_0^{p_B} p \cdot \alpha_0 \cdot dp + \frac{1}{g} \cdot \int_0^{p_B} p \cdot \delta \cdot dp = E_p^0 + \chi . \tag{9.27}$$

We have removed the minus sign using $\int_{p_B}^{0} = -\int_{0}^{p_B}$. $E_p^o$ is a function of the
bottom pressure only; it is equal to the potential energy of sea water of
$S = 35\%_o$, $T = 0°C$. $\chi$ is the potential energy anomaly – the difference from
$E_p^o$ associated with the difference between the reference specific volume and
the actual specific volume; it is sometimes a useful function to consider in
synoptic physical oceanography. This separation is similar to the one made
when discussing the geopotential in the previous chapter and $\chi$ can be calculated from oceanographic observations in a similar way to that used to obtain
$\Delta\Phi$. Now consider $\partial E_p/\partial x$ :

$$\frac{\partial}{\partial x}\int_{z_B}^{\eta} p \cdot dz = \int_{z_B}^{\eta} \frac{\partial p}{\partial x} \cdot dz + p_\eta \cdot \frac{\partial \eta}{\partial x} - p_B \cdot \frac{\partial z_B}{\partial x}$$

The terms $p_\eta \cdot (\partial \eta/\partial x)$ and $p_B \cdot (\partial z_B/\partial x)$ occur because $\eta$ and $z_B$ vary with x.
Since we take $p_\eta = 0$, the first of these vanishes. (It can be retained if
important, e.g., in storm surge calculations, but is not normally important
for the large-scale, steady-state circulation.) Rearranging:

$$\int_{z_B}^{\eta} \frac{\partial p}{\partial x} \cdot dz = \frac{\partial E_p}{\partial x} + p_B \cdot \frac{\partial z_B}{\partial x}$$

$$= \frac{\partial \chi}{\partial x} + \frac{\partial E_p^o}{\partial x} + p_B \cdot \frac{\partial z_B}{\partial x} \quad . \tag{9.28}$$

The last two terms may be combined (see Fofonoff) to give:

$$\int_{z_B}^{\eta} \frac{\partial p}{\partial x} \cdot dz = \frac{\partial \chi}{\partial x} + \frac{p_B \cdot \alpha_B}{g} \cdot \left(\frac{\partial p_B}{\partial x}\right)_z \tag{9.29}$$

where $(\partial p_B/\partial x)_z$ means the change of pressure on a level surface at the bottom.
The same equation will hold with y replacing x, so we can substitute these
equations in 9.26 to give:

$$- f \cdot M_y = -\frac{\partial \chi}{\partial x} - \frac{p_B \cdot \alpha_B}{g} \cdot \left(\frac{\partial p_B}{\partial x}\right)_z + \tau_{x\eta} - \tau_{xB} \quad ,$$

$$\tag{9.30}$$

$$f \cdot M_x = -\frac{\partial \chi}{\partial y} - \frac{p_B \cdot \alpha_B}{g} \cdot \left(\frac{\partial p_B}{\partial y}\right)_z + \tau_{y\eta} - \tau_{yB} \quad .$$

The first terms are associated with variations in the density distribution;
they arise from the baroclinic part of the geostrophic velocity, i.e.,
$f \cdot M_{yc} = \frac{\partial \chi}{\partial x}$, $f \cdot M_{xc} = -\frac{\partial \chi}{\partial y}$. The stress terms may be associated with Ekman
transports as we did earlier:

$$f \cdot M_{yE} = -\tau_{x\eta}; \quad f \cdot M_{yEB} = \tau_{xB}; \quad f \cdot M_{xE} = \tau_{y\eta}; \quad f \cdot M_{xEB} = -\tau_{yB} \quad . \quad \text{The}$$

terms involving bottom pressure are the barotropic transports times f.

$$M_{yb} = \int_{z_B}^{\eta} \rho \cdot v_b \cdot dz = v_b \cdot \int_{z_B}^{\eta} \rho \cdot dz \text{ because } v_b \text{ does not vary with } z.$$

Now $\rho \cdot dz = - dp/g$, so $\int_{z_B}^{\eta} \rho \cdot dz = - \int_{p_B}^{o} dp/g = p_B/g$, and $M_{yb} = \dfrac{v_b \cdot p_B}{g}$ .

The barotropic flow is the geostrophic flow in the deep water, so

$$f \cdot v_b = \alpha_B \cdot \left(\frac{\partial p_B}{\partial x}\right)_z \quad \text{and} \quad \frac{p_B \cdot \alpha_B}{g} \cdot \left(\frac{\partial p_B}{\partial x}\right)_z = \frac{f \cdot v_b \cdot p_B}{g} = f \cdot M_{yb} \quad .$$

Likewise, the $p_B$ term in the second equation of 9.30 is $- f \cdot M_{xb}$. Thus equation 9.30 may be written as:

$$M_y = M_{yc} + M_{yb} + M_{yE} + M_{yEB}$$

$$M_x = M_{xc} + M_{xb} + M_{xE} + M_{xEB} \quad .$$

(9.30')

These equations simply divide the total transport into components which are related to the density field, bottom pressure gradient and stress terms of equation 9.30.

Now, following Sverdrup's approach we take

$$\frac{\partial}{\partial x} (f \cdot M_x) + \frac{\partial}{\partial y} (f \cdot M_y) = f \cdot \left(\frac{\partial M_x}{\partial x} + \frac{\partial M_y}{\partial y}\right) + \beta \cdot M_y = \beta \cdot M_y \quad (9.31)$$

because by continuity $(\partial M_x/\partial x + \partial M_y/\partial y) = 0$ for the total transport from surface to bottom. Following the same procedure for the right-hand sides of equations 9.30 gives:

$$\beta \cdot M_y = \text{curl}_z \, \underline{\tau}_\eta - \text{curl}_z \, \underline{\tau}_B - \rho' \cdot f \cdot \left(u_b \cdot \frac{\partial z_B}{\partial x} + v_b \cdot \frac{\partial z_B}{\partial y}\right) \quad (9.32)$$

or in vector form;

$$\beta \cdot M_y = (\nabla \times \underline{\tau}_\eta)_k - (\nabla \times \underline{\tau}_B)_k - \rho' \cdot f \cdot \underline{V}_b \cdot \nabla_H z_B$$

where $\nabla_H = \underline{i} \cdot (\partial/\partial x) + \underline{j} \cdot (\partial/\partial y)$ and $\rho' = \rho_B \cdot \left[1 + \dfrac{p_B}{\alpha_B}\left(\dfrac{\partial \alpha}{\partial p}\right)_B\right]$ (see Fofonoff for a partial derivation of the last term). The $\chi$ terms cancel out just as did the pressure gradient terms in Sverdrup's derivation because he assumed the deep flow (barotropic part) to be zero. With no flow near the bottom there is no stress there either, and 9.32 reduces to the simpler form of 9.21 derived by Sverdrup. If the bottom is level and the bottom stress negligible one also gets the simpler form (9.21). Finally, if the flow is along the bottom contours, i.e., is entirely horizontal so that $\underline{V}_b$ is perpendicular to $\nabla_H z_B$, then the terms involving $z_B$ also vanish. It has been quite common to neglect the bottom stress; this approximation is probably good in most cases.

It has also been quite common to neglect the $\underline{V}_b \cdot \nabla_H z_B$ term based on the idea that the velocities vanish at great depths since the driving for the flow is

from the surface.  It is only possible to include this effect in analytic
treatments (with suitably idealized bottom topography) if the baroclinicity is
also excluded, i.e., the geostrophic part of the flow is depth independent,
another reason for leaving it out.  Even if we take the density field as given
from observations, the difficulty remains that we cannot obtain $V_b$ observation-
ally because we cannot obtain it from geostrophic calculations, and deep
current meter observations are too few and of too short duration to help much.
As we shall see in Chapter II on numerical models, the $\underline{V}_b \cdot \nabla_H z_B$ term may be
important in the real ocean.

### THE MASS TRANSPORT STREAM FUNCTION

By integrating along the vertical we have produced the mass transport per unit
width of current (the density times the vertically averaged velocity times the
depth) which depends on x and y but not on z.  When we have such a flow,
which depends on only two space variables (and is either incompressible or
steady state), it is possible to use a scalar function called a *stream function*
from which the velocity may be derived.  This approach can provide a useful
simplification because it may be easier to find a single scalar function than
the two components of a vector.  We put

$$M_x = \frac{\partial \psi}{\partial y}, \quad M_y = -\frac{\partial \psi}{\partial x} \qquad\qquad (9.33)$$

where $\psi$ is the stream function.  The flow is parallel to lines on which $\psi$ is a
constant, which means that plots showing such lines are convenient for display
purposes as mentioned earlier in connection with Fig. 9.11 and described
somewhat more fully in Appendix I.*

Consider $\quad \dfrac{\partial M_x}{\partial x} + \dfrac{\partial M_y}{\partial y} = \dfrac{\partial^2 \psi}{\partial x \partial y} - \dfrac{\partial^2 \psi}{\partial y \partial x}$ .

Now by continuity, if we take the total transport this quantity should vanish
and the order of differentiation of $\psi$ can be interchanged (which is a mathematical condition for $\psi$ to be well behaved, as the interested reader may see
by consulting a book on fluid mechanics, e.g., Batchelor).  In practice, if
we can find an equation for $\psi$ from the equations of motion, usually in the
form of the Sverdrup equation 9.21, or equation 9.32 the more general form, or
extensions discussed presently, the solution will automatically be such that
continuity is satisfied.  To get an equation for $\psi$ in the case where the
Sverdrup equation holds we simply replace the $\beta \cdot M_y$ term by $-\beta \cdot (\partial \psi / \partial x)$.

---

*
The reader should note that some writers introduce the minus sign in the $M_x$
equation of 9.33 - we have chosen to follow what we believe is the more common
practice in fluid mechanics although not, unfortunately, in physical oceanography which is rather divided on this matter.  If the opposite sign convention to ours is used and a $\psi = 0$ streamline is used, as is common, the signs
of the $\psi$ values in our convention must be multiplied by -1 to convert.  Normally there is no problem in interpreting the plots because arrows are usually
placed on the lines to show the direction of flow.

**Wind stress**

Fig. 9.12.  Flow patterns (streamlines) for simplified wind-driven circulation
with: (a) Coriolis force zero or constant, (b) Coriolis force
increasing linearly with latitude (from Stommel, *Trans. A.G.U.*,
29, 205, 1948).

While the total transport is non-divergent (i.e.,continuity takes the form
$(\partial M_x/\partial x)+(\partial M_y/\partial y) = \nabla_H \cdot \underline{M} = 0)$ the individual transports, Ekman, Baroclinic,
Barotropic (if any) and Bottom Ekman (if any) may not be so; it may not be
possible to represent them by stream functions.  The negative of the potential
energy anomaly $(-\chi)$ is almost a stream function for the baroclinic transport.
Actually, contours of $(-\chi)$ are 'streamlines' for $f \cdot \underline{M}_C$ but since f varies
slowly the baroclinic flow will almost be along the contours.  The relation
between the contour spacing and the strength of the transport will vary with
f and hence with latitude.

Similarly, a contour plot of $\Delta\Phi$ (or $\Delta D$ in the mixed units system) shows the
pattern of the horizontal geostrophic flow (relative to that at the reference
level).  It is not a true stream function unless $(\partial u/\partial x)+(\partial v/\partial y) = 0$ is also
true, but it provides the same useful display features as a stream function.


## WESTWARD INTENSIFICATION - STOMMEL'S CONTRIBUTION

A feature of the ocean circulation seen as a whole is the so-called *westward
intensification*, for example as shown in Fig. A.3 (Appendix I) for the North
Pacific where the flow lines are close together in the west or north-west off
Japan, whereas they are more widely separated over most of the rest of the
ocean.  Where the flow lines are close, the flow must be swift, and *vice versa*.
Similar flow patterns are evident in the North and South Atlantic, probably in
the Indian Ocean, but less evident in the South Pacific.  (In the latter case
the flow is complicated by the islands in the west.)  Stommel was the first
to present an adequate explanation for this feature.

His demonstration was done with a simplified theoretical model of an ocean and
wind pattern.  He took a rectangular ocean of constant depth and all on one
side of the equator, and assumed the earth to be flat for convenience, i.e.,
he used the tangent plane approximation mentioned in Chapter 6.  He also
assumed a wind stress which varied with latitude as shown in Fig. 9.12, so
that it was to the west at the south and to the east at the north.  This is a

Fig. 9.13.    General pattern of global winds and mean E-W wind components.


reasonable approximation to the real wind stress which will be discussed
shortly.  He also included a simple friction term to prevent acceleration so
that he could investigate the steady state.  Basically he used Sverdrup's
equation (9.21) with friction added.  He calculated the flow patterns in this
ocean for three conditions:

(a)    non-rotating ocean (i.e., non-rotating earth),
(b)    rotating ocean but Coriolis parameter f constant (the f-plane
       approximation),
(c)    rotating ocean with Coriolis parameter varying with latitude
       φ in a simple but realistic fashion, i.e., linearly with φ
       from 10° to 50° latitude (the β-plane approximation).

The flow patterns which he obtained then appeared as in Fig. 9.12, the second
being the most like the flow pattern in the real oceans (e.g., as in Fig.
A.3).  In case (a) the surface remained nearly level.  In case (b), to balance
the Coriolis effect a higher water level was found at the centre to provide
the necessary pressure gradient.  A similar high level was found in case (c)
but it was not symmetric in the east-west direction as in case (b).

It was therefore clear that the variation with latitude, $\partial f/\partial \phi$, of the Coriolis
parameter was responsible for the westward intensification.  Nowadays, this
result is discussed in terms of vorticity as will be described shortly.
(Basically, wind stress puts vorticity, i.e., spin, into the ocean and friction
is required to take it out.  When f varies with latitude, strong friction in
the west is needed to take out the vorticity and for this strong friction to
occur strong currents with strong shear are needed.)

This result of Stommel's was obtained for a very simplified version of the
real ocean, but it is clear that the variation of the Coriolis parameter is a
fundamental feature of the dynamics which must be taken into account in any

Fig. 9.14.    Form of drag coefficient ($C_D$) for wind over water as a function
of wind speed, as used for many calculations of wind-driven
circulation.


large ocean circulations.  In addition, Stommel's addition of a friction term
permitted a solution with a closed circulation, which Sverdrup's assumptions
did not.  Stommel's model was not intended to represent a real ocean but to
illustrate a principle, as is often the case with models which are much idealized from the real world.  The mathematical details are not important, provided that one believes the result (as we do in this case), and we have not
given them here.  The interested reader may consult Stommel's book (1964)
which also contains much other interesting information.


### THE PLANETARY WIND FIELD

Having indicated the development of the wind stress explanation for the upper
layer currents we should look for a few moments at the general character of
the winds on a global scale.  The main features are shown in Fig. 9.13.  On
the left are shown the main wind systems which are distributed in a zonal
fashion.  The graph of the east-west component of the wind on the right shows
that Stommel's form for the wind stress as a function of latitude is a reasonable approximation to the real stress between about 10° and 50°N.

This graph is an average across the whole width of the ocean, i.e., east-west
variations are ignored.  In more recent treatments, particularly the numerical
ones (Chap. 11), values for the wind stress for a grid of points over the
ocean are used so that a better approximation to the real wind stress both
north-south and east-west is applied.  Of course, it is still a time-averaged
distribution.

The presently available wind stress results are probably not very representative in detail.  The procedure is to use the relation that wind stress
$\tau = \rho \cdot C_D \cdot W^2$ in the direction of the wind $\underline{W}$.  One problem is in the value to
use for the drag coefficient $C_D$.  As far as we know, most stress calculations

for oceanographic purposes have used a step or smoothed step function for $C_D$ as in Fig. 9.14. More recent and more accurate measurements suggest that $C_D = 1.3$ to $1.5 \times 10^{-3}$ for wind speeds up to about $15\,\mathrm{m\,s^{-1}}$ is better, with perhaps a slight tendency to increase over this speed range. Recently, new measurements at wind speeds up to $20\,\mathrm{m\,s^{-1}}$ or so have been obtained but analysis of these results is still in progress. Once analysis is completed, calculations of wind stress over the oceans should be better because winds above $20\,\mathrm{m\,s^{-1}}$ are too infrequent to make much contribution to the average stress in most regions of the ocean. (Of course, very strong winds do have dramatic effects locally while they last and accurate determination of their effects will still be uncertain until direct measurements are obtained.)

The presently available indirect measurements at high wind speeds on which the step or smoothed step function is based are now considered to be unreliable. They were based on 'set-up' of the water level, i.e., rise above predicted tidal level. Unfortunately the problem is more complicated than a simple balance between the pressure gradient associated with the set-up and the wind stress at the surface. Time variations, the effect of bottom topography and non-linear effects in shallow water on the waves associated with the strong winds where the set-up measurements were made are all important factors. The step function for $C_D$ is based on the apparent large values of $C_D$ from such measurements and the lower values at light wind speeds from more direct measurements obtained initially. These low values may be too low because it is the difference in air-water velocity that is important. Under light winds, a relatively thin layer of water may be moving relatively quickly due to the wind and the value of W used to calculate $C_D$ may be too large. The jump of $C_D$ as shown in Fig. 9.14 was rationalized on the basis that wave breaking or 'white capping' becomes *obvious* at winds of about $7.5\,\mathrm{m\,s^{-1}}$ (15 knots) and the water surface becomes 'rougher' leading to higher values for $C_D$.

The use of the step function persisted long past the time when it was known not to be realistic. This provides an example of the unfortunate use, by people not fully conversant with the development of a particular field, of preliminary results as though they were settled matters.

Another difficulty is that the wind speed appears in the stress expression as a power of W. If we have frequent measurements of $\underline{W}$, e.g., daily or more frequently) we can calculate the stress for each measurement and add these *stress* values vectorially to obtain the total effect. Unfortunately, in the early calculations mean values ($\overline{W}$) for the wind speed over periods of a month or more were used. If the values of W change much during the averaging period, the square of this mean value [i.e., $(\overline{W})^2$] is not the same as the mean of the squared values [i.e., $(\overline{W^2})$]. The relation between the two is: $(\overline{W^2}) = (\overline{W})^2 + \sigma_w^2$ where $\sigma_w^2$ is the variance of the values of W. The correction is not negligible. For instance, if $\sigma_w = 0.5W$, not unlikely in mid-latitudes if averages of a month are used, then the quantity which we need for the wind stress $(\overline{W^2}) = 1.25\,(\overline{W})^2$. As a simple example, if the wind were to blow at $5\,\mathrm{m\,s^{-1}}$ for 9 days and then at $15\,\mathrm{m\,s^{-1}}$ for one day, the relative values of $\Sigma[(\overline{W^2}) \times t]$ to $[(\overline{W})^2 \times t]$ are in the ratio of 1.25 to 1, i.e., the use of long-term mean values tends to underestimate the effect of the wind stress. Another way to put it is to point out that for momentum transfer (i.e., frictional stress), for a constant drag coefficient, 1 day at $25\,\mathrm{m\,s^{-1}}$ is equivalent to about 1 week at $10\,\mathrm{m\,s^{-1}}$ or about 1 month at $5\,\mathrm{m\,s^{-1}}$.

Variability in mid-latitudes is rather greater than in the trade wind regions and cannot be neglected. Climatological wind data are summarized in the form of wind roses which show the percentage of the time for which the wind blows in particular directions (usually eight) in each of a number of speed ranges, as well as the overall frequency distribution of speeds regardless of direction.  Using wind roses one can calculate the contribution to the stress from each direction in each speed range and add them vectorially to obtain the climatological average *stress*.  A few tests, using detailed wind data to calculate the stress both from the detailed data and from the same data put into wind rose form, show that both methods give the same results within a few percent.  Thus once $C_D$ is better known (hopefully soon) calculations of the climatological average stress over the ocean should be possible with reasonable accuracy.

## MUNK'S SOLUTION

Munk combined the basic features contributed by Ekman, Sverdrup and Stommel to provide the first comprehensive solution of the wind-driven circulation, using the real wind field, albeit with less detailed wind roses than presently available and with the step function $C_D$.  He used two friction terms:

(a)    vertical, associated with vertical shear to convey momentum from the wind stress applied at the surface into the Ekman layer,

(b)    lateral, associated with horizontal shear so that the ocean would remain in a steady state of circulation.

He finished up with a fourth order differential equation describing the circulation as:

$$A \cdot \nabla^4 \psi - \left( \beta \cdot \frac{\partial \psi}{\partial x} \right) - \mathrm{curl}_z\, \underline{\tau}_\eta = 0 \qquad\qquad (9.34)$$

where    A    =    the eddy viscosity coefficient for lateral friction for mass transports,
$\nabla^4$    =    the two-dimensional biharmonic operator $= \dfrac{\partial^4}{\partial x^4} + 2 \cdot \dfrac{\partial^4}{\partial x^2 \partial y^2} + \dfrac{\partial^4}{\partial y^4}$,
$\psi$    =    the mass transport stream function which describes the stream lines (really trajectories in the steady state) of flow around the ocean.

The equation must be solved for $\psi$ and then:

$$M_x = \frac{\partial \psi}{\partial y}, \qquad M_y = -\frac{\partial \psi}{\partial x}.$$

In words, equation 9.34 can be written:

   Vorticity from lateral stress − Planetary vorticity − Wind stress curl = 0.

(Vorticity will be discussed shortly.)

The three terms in the equation are not equally important all over the ocean. In the west, where the currents are strong, the first and second are the important ones while in the remainder of the ocean the second and third are important.  The lateral stress is determined by the lateral shear in the currents and is large in the west because the currents and shear are large there.  Elsewhere the currents are so much less that the terms arising from the shear can never be large as we showed in Chapter 7.

Munk solved the differential equation, i.e., obtained an expression for $\psi$ in terms of the dimensions of the ocean, of $\beta$ and of $\tau_n$. For the latter he used values for the east-west wind stress component (averaged across the ocean) only, ignoring the north-south component. For the solution, $\psi$ is best shown in the form of the flow pattern as in Fig. 9.15a. This is the solution for a rectangular ocean and is somewhat stylised. Later, Munk and Carrier solved for a triangular ocean which more nearly represents the real shape of the North Pacific as shown in Fig. 9.15b.

Note that the second and third terms are just the Sverdrup equation 9.21 with $-\partial\psi/\partial x$ substituted for $M_y$. The additional terms in the generalized Sverdrup equation 9.32 do not appear because Munk assumed that the currents went to zero above the bottom or that the bottom was level and the bottom stress was negligible because the currents ought to be very small in the deep water.

Munk assumed that in the west the friction terms associated with horizontal shears in the currents would become important but that non-linear terms would remain small. Near the end of Chapter 8 we showed that this assumption may be reasonable if the horizontal eddy viscosity is sufficiently large.

To see how Munks' equation 9.34 is obtained we write down the vertically integrated equations 9.30 but add the lateral friction terms from equation 7.6. For simplicity we omit the bottom pressure and stress terms as under Munk's assumptions they are taken to be zero:

$$- f \cdot M_y \;=\; - \frac{\partial \chi}{\partial x} + \tau_{x\eta} + \int_{z_B}^{\eta} \rho \cdot A_x \cdot \frac{\partial^2 u}{\partial x^2} \cdot dz + \int_{z_B}^{\eta} \rho \cdot A_y \cdot \frac{\partial^2 u}{\partial y^2} \cdot dz$$

$$f \cdot M_x \;=\; - \frac{\partial \chi}{\partial y} + \tau_{y\eta} + \int_{z_B}^{\eta} \rho \cdot A_x \cdot \frac{\partial^2 v}{\partial x^2} \cdot dz + \int_{z_B}^{\eta} \rho \cdot A_y \cdot \frac{\partial^2 v}{\partial y^2} \cdot dz \;.$$

Now assume that $A_x = A_y = A_H$ and that $\int_{z_B}^{\eta} \rho \cdot A_H \cdot \frac{\partial^2 u}{\partial x^2} \cdot dz \simeq A \cdot \frac{\partial^2 M_x}{\partial x^2}$ and

likewise for the other terms. Now if $z_B$ = constant as Munk assumed and $A_H$ does not vary with z, the result is nearly exact because the $\eta$ variations are too small to matter and density variations can also be ignored in this case (by the Boussinesq approximation discussed in Chapter 7). Recall that we have already assumed that $A_x$ and $A_y$ variations with x and y may be ignored. If some of the assumptions are not exactly correct then we argue by analogy that friction for transports may be represented by $A \cdot (\partial^2 M_x / \partial x^2)$, etc., with $A \simeq A_H$. Recall also, however, that the use of the eddy viscosity, particularly a constant value, is a crude way to represent the effects of turbulence. Any results which depend strongly on this assumption must be viewed with suspicion until verified by observations.

Using the eddy viscosity representation gives friction terms $A \cdot \nabla_H^2 M_x$ in the first equation and $A \cdot \nabla_H^2 M_y$ in the second (where $\nabla_H^2 = \partial^2/\partial x^2 + \partial^2/\partial y^2$). Now we take $[\partial(f \cdot M_x)/\partial x + \partial(f \cdot M_y)/\partial y]$ as we did earlier. Except for the friction term we already know the result ($\beta \cdot M_y = \operatorname{curl}_z \tau_\eta$) so we just work out its form assuming that A is constant:

$$A \cdot \frac{\partial}{\partial x} (\nabla_H^2 M_y) \;-\; A \cdot \frac{\partial}{\partial y} (\nabla_H^2 M_x) \;.$$

Fig. 9.15.  (a) on left - mean annual wind stress over the Pacific, $\tau_x$ (full
           line) and its curl, $10^8 \partial\tau_x/\partial y$ (dashed line); on right - computed
           mass transport streamlines ($\psi$) for a rectangular ocean (from
           W.H. Munk, *J. Met.*, 7, 82, 1950), (b) computed transport stream-
           lines for a triangular ocean from 15° to 60°N (from W.H. Munk &
           G.F. Carrier, *Tellus*, 2, 163, 1950). (1 dyne cm$^{-2}$ ≡ 0.1 Pa;
           1 dyne cm$^{-3}$ ≡ 10 Pa m$^{-1}$.)

Using $M_x = \partial\psi/\partial y$, $M_y = -\partial\psi/\partial x$ and writing out in full we have:

$$- A \cdot \left[ \frac{\partial}{\partial x} \left( \frac{\partial^2}{\partial x^2} + \frac{\partial^2}{\partial y^2} \right) \cdot \frac{\partial\psi}{\partial x} + \frac{\partial}{\partial y} \left( \frac{\partial^2}{\partial x^2} + \frac{\partial^2}{\partial y^2} \right) \cdot \frac{\partial\psi}{\partial y} \right]$$

$$= - A \cdot \left[ \frac{\partial^4\psi}{\partial x^4} + 2 \cdot \frac{\partial^4\psi}{\partial x^2 \partial y^2} + \frac{\partial^4\psi}{\partial y^4} \right] = - A \cdot \nabla^4\psi$$

and we have finally:

$$\beta \cdot M_y = - \beta \cdot \frac{\partial\psi}{\partial x} = \mathrm{curl}_z \, \underline{\tau}_\eta - A \cdot \nabla^4\psi$$

which is Munk's equation.

Note that this is a fourth order equation (i.e., it contains fourth partial derivatives) and therefore its solution can satisfy four boundary conditions – no flow through and no-slip along both east and west boundaries. The vanishing of $\mathrm{curl}_z \, \underline{\tau}_\eta$ at certain latitudes breaks the flow into gyres as in Fig. 9.15. In Stommel's model, because of the simpler form which he assumed for the friction term, his equation was only of second order and he could not satisfy the no-slip condition. Also because of the higher order, Munk's solution allows for the counter current (a fairly strong southward flow observed to the east of the Kuroshio and the Gulf Stream). In Fig. 9.15, the western boundary current is where the streamlines are close together indicating strong currents. The counter current is indicated in the largest gyre by the swing to the south of the streamlines as they enter and leave this region. Stommel's model (Fig. 9.12b) does show the western intensification but not the counter current. Munk's solution is more realistic in these regards.

Comments on Munk's Solution

These solutions show a series of 'gyres' which include the equatorial current system and the westward intensification. Quantitatively, for the larger currents such as the Gulf Stream and the Kuroshio, Munk's calculated values for the transports are only about one-half of the observed values from geostrophic calculations. The calculated transports are based on integrating $M_y$ obtained from the Sverdrup equation (9.21) across the ocean in the x-direction and finding the maximum value. Thus these values do not depend on the eddy viscosity at all but only on the validity of equation 9.21. The geostrophic calculations are done across the boundary currents themselves and have uncertainty because the level of no motion is uncertain. Direct observations of the transport of the Gulf Stream suggest even larger disagreement. Part of this disagreement may be due to an underestimate of the wind stress and more particularly its curl which is estimated using finite differences. The stress used is usually calculated at 5° intervals of latitude and longitude and the curl may be underestimated using such large separations, particularly where it has maxima and minima.

In addition, there is the doubt about the value to use for $C_D$. Munk noted that using $C_D = 2.6 \times 10^{-3}$ everywhere would give better results. This value seems high, particularly in view of more recent estimates, but the winds which are used are somewhat averaged. Also, the observations are from ships'

reports and merchant and passenger ships try to avoid strong winds, leading
to some bias toward low wind-speeds in the data and consequent underestimates
of the stress.  Using a larger $C_D$ would tend to correct the underestimate but
a factor of almost 2 larger seems excessive.

The wind stress estimates are based on wind data over many years, the curl
being estimated from these stresses.  In addition to the possible error because the finite differences used are based on large separations, the locations of the regions of maxima and minima of wind stress curl will vary
seasonally and also from year to year in a given season.  Thus the maxima and
minima of north-south transport integrated across the ocean calculated from
climatological data will be of smaller magnitude than the values which one
would obtain by averaging the maxima or minima of the integrated transport
(regardless of latitude) for a particular gyre based on daily values of the
stress.  How important this effect might be is unknown because, as far as we
know, such calculations have not been made.  Overall, a factor of two effect
is probably an upper limit.  Daily stress values could only be estimated by
extrapolating the geostrophic wind from surface pressure maps to a surface
wind.  To do such a calculation for a few years would be a large task.  Also
there is uncertainty both in the maps and in the extrapolation.  However,
such a calculation over the range of latitudes in which the largest southward
integrated transport in the Gulf Stream gyre is likely could prove to be
quite useful.

In summary, it appears that there is a real discrepancy between the calculated
purely wind-driven ocean circulation and that in the real ocean.

Munk neglects the thermohaline circulation entirely but does not think that it
is a significant source of error.  Stommel has suggested that the thermohaline
circulation may provide a significant contribution to the total flow.  Furthermore, the neglect of the non-linear terms in the acceleration, such as
$u \cdot (\partial u/\partial x) + v \cdot (\partial u/\partial y) + w \cdot (\partial u/\partial z)$, may not be justifiable in the western
boundary current region.  Including these terms makes the equations non-linear
but attempts have been made by Morgan and by Charney to develop analytic
theories in which the inertial terms are included.  No fully satisfactory
theory has yet developed but it appears that the inertial or non-linear terms
must be taken into account in some circumstances and may then be just as
important in the west as the lateral friction term.  More recent attempts at
calculating the ocean circulation have been made with numerical models in
which all the effects, including the inertial ones, can be included.  Friction
is still approximated with constant eddy viscosity in most cases although
other formulations are possible (e.g., larger values where velocity gradients
are larger which, from what we know of turbulent motion, seems more realistic).
These more 'sophisticated' eddy viscosities have been used more in models of
the atmosphere which are somewhat more advanced than models of the ocean.
Numerical models have shown that inertial effects may increase the boundary
current transport above that of the interior Sverdrup transport (i.e., that
calculated using the Sverdrup equation 9.21).  The effect of bottom topography
in a realistic baroclinic ocean according to some model results may also be
responsible for the enhanced boundary current  flow (in the sense of being
larger than predicted by the simple wind theory just described).  Observations
of salinity and temperature in the deep water are not accurate enough to show
clearly whether this mechanism is really important.  There are other possible
enhancement mechanisms also.  Features of numerical models will be discussed
in Chapter II.

Fig. 9.16.    For discussion of vorticity: (a) relation between relative
              vorticity ($\zeta$) and velocity shear, (b) planetary vorticity (f)
              at various latitudes on a rotating earth.


VORTICITY

Relative Vorticity ($\zeta$)

Expressed simply, *vorticity* is a characteristic of the kinematics of fluid
flow which expresses the tendency for portions of the fluid to rotate.  It is
directly associated with the quantity called 'velocity shear'.  To illustrate
this relationship, Fig. 9.16 shows on the left a plan view of some fluid
which is flowing to the right (east) with velocity u(y), the variation with y
being such that from top to bottom of the figure, (A) the velocity first

Fig. 9.17.    Change of absolute vorticity associated with (a) convergence, (b) divergence.

increases, then (B) is constant for a space and then (C) decreases. A small object floating in the water in zone A would tend to rotate in an anticlockwise direction as it drifted to the right as shown successively for times $t = t_1, t_2, t_3$, etc. An object in zone C would tend to rotate clockwise, and one in the centre zone B would not tend to rotate either way. The rotation of the fluid, in this case measured by $\partial u/\partial y$, is called the vorticity. When it is measured relative to the earth it is called *relative vorticity* ($\zeta$, zeta). When it is measured relative to axes fixed in space it is called 'absolute vorticity' (discussed later).

The sign convention for direction is that the vorticity is positive when it is anticlockwise (zone A) as viewed from above (the same direction as the rotation of the earth as viewed from above the North Pole), and negative when clockwise (zone C).

In the general case, the relative vorticity in the horizontal plane (the vertical component) is $\zeta = \mathrm{curl}_z \underline{V} = (\partial v/\partial x - \partial u/\partial y)$.

Planetary Vorticity (f)

For a rotating solid object, the vorticity = 2 x angular velocity. By virtue of the rotation of the earth in space, at latitude $\phi$ a portion of its surface has angular velocity $\Omega \cdot \sin\phi$ about a vertical axis and therefore vorticity $2\Omega \cdot \sin\phi$. This is called *planetary vorticity*. It is the quantity f appearing in some of the Coriolis terms and we will continue to use the symbol

$f = 2\Omega \cdot \sin\phi$. A body of water which is stationary relative to the earth will then automatically possess planetary vorticity $f$; Fig. 9.16b shows how this quantity varies with position on the surface of the earth. Note that the planetary vorticity varies only with latitude $\phi$ and therefore a parcel of water at latitude $\phi_1$ can have only $f_1 = 2\Omega \cdot \sin\phi_1$, no other value is possible. Also notice that f is zero at the equator, increasing to $+2\Omega$ at the north pole and decreasing to $-2\Omega$ at the South Pole.

## Absolute Vorticity $(\zeta + f)$

The equations for the horizontal components of motion without friction are:

$$\frac{du}{dt} - f \cdot v = -\alpha \cdot \frac{\partial p}{\partial x}$$

$$\frac{dv}{dt} + f \cdot u = -\alpha \cdot \frac{\partial p}{\partial y}$$

$$(9.35)$$

If we cross-differentiate these equations and subtract them, to eliminate the pressure terms, we get:

$$\frac{d}{dt}(\zeta + f) = -(\zeta + f) \cdot \left[\frac{\partial u}{\partial x} + \frac{\partial v}{\partial y}\right] = -(\zeta + f) \cdot \nabla \cdot \underline{V}_H \qquad (9.36)$$

where $\underline{V}_H$ stands for the horizontal velocity and $\nabla \cdot \underline{V}_H$ will be recognized as a measure of the tendency for horizontal flow to diverge or converge. The quantity $(\zeta + f)$, the sum of the relative and planetary vorticities, is called the *absolute vorticity*. Equation 9.36 expresses the Principle of Conservation of Absolute Vorticity for flows on the earth when frictional effects are neglected. Here we have neglected derivatives of $\alpha$ with respect to x and y as usual. Consistent with this approximation, terms involving vertical shear are neglected. (In the ocean, for vorticity one may treat the flow as barotropic, taking u, v and $\zeta$ as independent of z as a good approximation.) Note also that $df/dt = v \cdot (\partial f/\partial y) = \beta \cdot v$ since f is independent of x, z and t.

In a divergence, where $\nabla \cdot \underline{V}_H$ is positive, the magnitude of the absolute vorticity decreases with time, whereas in a convergence where $\nabla \cdot \underline{V}_H$ is negative, the absolute vorticity magnitude increases with time. We consider the magnitude because $(\zeta + f)$ may be positive or negative. As f is usually much larger than $\zeta$, positive values for $(\zeta + f)$ will usually be found in the northern hemisphere and *vice versa*. With $\nabla \cdot \underline{V}_H > 0$, if $(\zeta + f) > 0$ then $d(\zeta + f)/dt < 0$, so $(\zeta + f)$ decreases with time, but if $(\zeta + f) < 0$, $d(\zeta + f)/dt > 0$, i.e., $(\zeta + f)$ becomes more positive with time but as $(\zeta + f) < 0$ to start with, its magnitude decreases.

To get a physical picture of this process, imagine a body of water in the form of a vertical cylinder of small height which is initially stationary relative to the earth so that it has planetary vorticity f only, as in Fig. 9.17a. If the fluid now starts to flow inward (converges) toward the axis of the cylinder it must also elongate because volume is conserved; because $\nabla \cdot \underline{V}_H$ is negative, the absolute vorticity must increase. In this case the water will acquire some relative vorticity $\zeta$, so that its absolute vorticity increases from f to $(f + \zeta)$ as the cylinder shrinks and elongates. In Fig. 9.17b is shown the opposite situation in which a tall, narrow cylinder expands (diverges) to form a low, wide one. If its initial vorticity was simply f its final vorticity will decrease to $(f - |\zeta|)$. In both cases we have assumed that the cylinder of water remains at the same position on the earth's surface so that the planetary vorticity f does not change.  Note also that the curved arrows showing the motion of the water as it converges or diverges are drawn for the northern hemisphere case.

If one is more used to thinking in terms of conservation of angular momentum, then in the first case the moment of inertia is decreased; as the angular momentum (moment of inertia x angular velocity) is not changing, there being no applied torques in this case, the angular velocity must increase.  Similarly, in the second case the moment of inertia increases and the angular velocity decreases.

## Potential Vorticity $\left(\dfrac{\zeta + f}{D}\right)$

Let us consider a layer of thickness D in the sea whose density is uniform so that the horizontal velocity components are independent of depth.  (This thickness D is not specifically identified with the Ekman depth $D_E$.  D would more likely be the whole layer from the surface to the permanent thermocline or from the thermocline to the bottom.  This is an idealization of the real situation in the ocean to one of two homogeneous layers.  Clearly it is an approximation but the general features of the results will be correct and so it is often a useful one to make.  Then the equation of continuity of volume for the layer is:

$$\frac{1}{D} \frac{dD}{dt} + \left(\frac{\partial u}{\partial x} + \frac{\partial v}{\partial y}\right) = 0 \ . \tag{9.37}$$

If we combine this equation with 9.36, eliminating the horizontal divergence term, we get

$$\frac{d}{dt}\left(\frac{\zeta + f}{D}\right) = 0 \ , \ \text{i.e.,} \ \left(\frac{\zeta + f}{D}\right) = \text{constant} \tag{9.38}$$

for the motion of a water body in the ocean provided that there is no input of vorticity (such as might come from a wind stress or other frictional effects).  The quantity $(\zeta + f)/D$ is called the *potential vorticity* of the water.

This relation permits us to make some predictions about vorticity changes when a parcel of water moves from one place to another.  Let us consider some possibilities:

(a)  if D remains constant:

  (i)   then if a column of water moves zonally (along a parallel of latitude so that $\phi$ remains constant), then f remains constant and so must $\zeta$;

  (ii)  If a column of water moves meridionally (along a line of constant longitude) toward the north pole, then f automatically increases and $\zeta$ must decrease to keep $(\zeta + f)$ constant, i.e., the water acquires more negative (clockwise) rotation relative to the earth;

  (iii) conversely, if a column of water moves toward the south pole then it acquires more positive (anticlockwise) rotation.

(b)  if D increases, then $(\zeta + f)$ must increase if it is positive initially,

(i)   so if the water moves zonally, then f remains constant so $\zeta$
       must increase, i.e., the water acquires more positive(anti-
       clockwise) rotation;

(ii)  if the water moves meridionally toward the north pole, then
       f automatically increases, and it is not immediately obvious
       what $\zeta$ will do;

(iii) if the water moves meridionally toward the south pole, f de-
       creases and $\zeta$ must increase, i.e., the column acquires more
       positive (anticlockwise) rotation.

(c)  if D decreases, then $(\zeta + f)$ must decrease if it is initially positive.
      The reader is left to work out what will happen to $\zeta$ for zonal and for
      meridional flow in this case and also for D changes when $(\zeta + f)$ is
      initially negative.

In the interior of the ocean, for large scale processes, $\zeta$ is negligible compared with f.  In consequence, conservation of potential vorticity becomes
$f/D$ = constant.  For instance, if a water column stretches (D increases) then
f must increase in magnitude: the water must move toward the nearest pole,
north or south (because f is a function of latitude only), and *vice versa*.
One way in which a water column can stretch is for it to pass over a trough
in the bottom, or it can contract by passing over a ridge.  The condition
$f/D$ = constant permits us to predict which way a current will swing on passing
over bottom irregularities - equatorward over ridges and poleward over troughs
in both hemispheres.  The deflection of the flow required to keep $f/D$ constant
is sometimes termed *topographic steering*.

Note finally that several of the equations derived earlier in the chapter are
vertically integrated forms of the vorticity equation.  The Sverdrup equation
in both simple and general form is a vertically integrated form and includes
wind friction.  Stommel's equation (not given explicitly) and Munk's equation
are of the same form and include lateral friction as well as wind friction,
e.g., in equation 9.32 with v taken as independent of z, (a good approximation
as noted), $\beta \cdot M_y = \rho \cdot \beta \cdot v \cdot D = \rho \cdot D \cdot (df/dt)$; $D = \eta - z_B$ and variations in D
will be dominated by $z_B$ variations, so the final term of equation 9.32 is
$\rho' \cdot f \cdot (dD/dt)$.  Because non-linear terms were neglected, $\zeta$ terms do not
appear in equation 9.32; in equations 9.36 and 9.38 they arise from the non-
linear terms of equation 9.35.  The curl terms of equation 9.32 do not appear
in equations 9.36 and 9.38 because we left out friction.

## WESTWARD INTENSIFICATION OF OCEAN CURRENTS EXPLAINED USING CONSERVATION OF POTENTIAL VORTICITY

Let us consider a northern hemisphere ocean and assume, for simplicity, that
the depth D of the circulation is constant so that $(\zeta + f)$ = constant.  The
general character of the winds, to the west in the south and to the east in
the north, causes the upper layer circulation to be clockwise (i.e., the
angular rotation is negative.)  Then on the west side of the ocean the flow
will be to the north (Fig. 9.18a) and there will be a loss of relative vorticity (i.e., $-\zeta_p$) due to the northward movement which causes f to increase,
and a loss of relative vorticity ($-\zeta_\tau$) due to the wind stress (which provides
clockwise, i.e., negative vorticity).  There is, therefore, a net loss of
vorticity ($-\zeta_p - \zeta_\tau < 0$) on the west side.  On the east side (Fig. 9.18b) the
flow is to the south and so there is a gain of relative vorticity ($+\zeta_p$) due
to the decrease of f, and a loss of relative vorticity ($-\zeta_\tau$) due to the

Fig. 9.18.   Use of conservation of potential vorticity to explain westward
            intensification of ocean currents: (a) relative vorticity
            supplied by wind stress, (b) relative vorticity supplied by
            frictional stress at boundary.

wind stress.  Now Sverdrup demonstrated that on the east side of the ocean, $\beta \cdot M_y = \text{curl}_z \tau_n$ (equation 9.21), i.e., the gain and loss terms for relative vorticity will be about equal in the east and there will be no change during the southward flow, i.e., $+\zeta_p - \zeta_\tau \approx 0$ in the east.

Therefore, for the complete circulation it is necessary to supply vorticity to make up for that lost in the west, in order to keep the total vorticity constant.  The way to do so is through friction on the west side only, so that the loss of planetary vorticity and loss of vorticity due to the wind stress are made up by gain of vorticity from lateral shear in the water currents, i.e., $-\zeta_p - \zeta_\tau + \zeta_f \approx 0$.  For this balance we need a velocity struc- ture such as in Fig. 9.18b where there are strong currents with much shear in the west but slower currents with little shear in the east, i.e., westward intensification of the ocean currents.  Strong currents in the middle or east of the ocean do not introduce vorticity of the correct sign to provide a balance.  (Strong currents on the west and friction acting at the bottom would also work but this possibility seems less likely than lateral shear acting on the upper layer flow.)

Where the wind stress causes an anticlockwise circulation (the larger ones being in the southern hemisphere), similar arguments again lead to westward intensification.

Note that there are two essential features: f must be allowed to vary with latitude (as Stommel showed) and $\zeta$ must be small compared with f.  If the winds were very much stronger so that $\zeta$ became large compared with f, then f could be ignored and we would return to a symmetrical circulation with wind vorticity input balanced by some other frictional effect (lateral or bottom) throughout the whole ocean.  However, in the earth's oceans $\zeta \ll f$ for the large-scale flow and therefore westward intensification occurs.

## EQUATORIAL UNDERCURRENTS

Beneath the surface and embedded in the westward flowing Pacific Equatorial Current there exists a most remarkable current - the Equatorial Undercurrent. This current flows eastward, centred on the equator.  Maximum speeds are $1\,\text{m s}^{-1}$ or more at a depth of about 100 m; the total transport is estimated at 40 Sv.  The current is like a thin ribbon, about 0.2 km thick and 300 km wide.  It can be followed for some 14,000 km across most of the Pacific!  Its existence has been recognized for only 25 years, although once it was known to exist, evidence for it was found in earlier data.

A similar current exists in the Atlantic, and was actually first recognized in 1886 by Buchanan but his accounts were ignored.  There is also evidence for an equatorial undercurrent in the Indian Ocean during the north-east monsoon period (November-March). More complete descriptions of these currents may be found in Pickard (1975) or Neumann and Pierson (1966).

Before the observational discovery of these undercurrents there were no theo- retical predictions to suggest their existence.  How do they arise?  Consider first what is happening right at the equator where the Coriolis effect is zero.  The wind is blowing with a westerly component and the surface current is to the west.  Because there are land barriers at the western side the water tends to pile up there producing a surface slope up towards the west and consequently a pressure gradient force develops toward the east.  The surface layer will tend to be thicker in the west with the thermocline (and pycnocline) sloping opposite to the surface slope, i.e., upward to the east. At some considerable depth the opposite slope of the isopycnals relative to the surface can lead to baroclinic compensation and a reduction of the eastward pressure gradient to zero. At the surface the westward component of the wind stress can balance the eastward pressure gradient. As one goes below the surface, the stress caused by the surface wind decreases, particularly as one gets into the pycnocline where the strong static stability inhibits turbulence and downward transfer of westward momentum. The eastward pressure gradient can no longer be balanced by the downward momentum transfer and produces the eastward flowing undercurrent in the pycnocline region. Because of the strong eastward current there will be frictional forces (and perhaps inertial effects) which will balance the pressure gradient and allow a steady-state flow to occur.

As one goes away from the equator, Coriolis effects become important. In the surface layer with a westward directed wind, the Ekman transport will be away from the equator on both north and south sides so there will be a surface divergence (with consequent upwelling and increased biological production). Below the surface layer there must be convergence (flow toward the equator) to balance the surface divergence. North or south of about 1.5° latitude the Coriolis force associated with this equatorward flow can balance the eastward pressure gradient. (It is only very near to the equator that the Coriolis effect is so weak that the undercurrent is possible.)

Note that although the Coriolis force vanishes at the equator its variation with latitude is a maximum there and tends to stabilize the current. (The Coriolis factor = $2\Omega \cdot \sin\phi$ which tends to zero as $\phi \to 0°$ but $\partial(2\Omega \cdot \sin\phi)/\partial\phi = 2\Omega \cdot \cos\phi$ which tends to its maximum value of $2\Omega$ as $\phi \to 0°$.) If the eastward current wanders southward the Coriolis force will be to the left pushing it back toward the equator; if it goes northward the Coriolis force is to the right, again bringing it back toward the equator. In fact, the variation of the Coriolis force with latitude will tend to stabilize eastward flowing currents anywhere in the ocean and to accentuate meandering in westward flowing currents with the strongest effects at the equator where the variation is a maximum.

Evidence for north-south oscillations in the equatorial currents has been found. They are probably associated with the variation of the Coriolis effect with latitude and may be thought of as Rossby waves superimposed on the average current. These waves, which are possible because of the Coriolis effect variation, will be discussed in Chapter 12.

Note that the discussion above is not a complete theory of the undercurrent, by any means. However, it does suggest how such a current can occur and what physical effects must be included in a complete theory which should be able to predict, for example, the detailed velocity distribution in the current. Both horizontal and vertical friction terms would need to be included in the equations. Because of the strength of the current and its small vertical and lateral (north-south) scales, non-linear terms might well be important too. At present there is no accepted complete theory of the undercurrent. (The reader wanting more details could consult the literature. A paper by Philander in the Reviews of Geophysics and Space Physics, 11, 513-570, 1973, provides a fairly recent review of the subject.)

## THE BOUNDARY LAYER APPROACH

When we examined the sizes of the terms in the equations of motion in Chapter 7 we found that many terms were negligible. The equation for the vertical component became the hydrostatic equation; the local time derivatives were shown to be small for motions with periods greater than a few days. It appeared that friction from turbulence and non-linear terms might be important in some regions. If we divide through by the magnitude of the Coriolis term we find that the non-linear terms and friction terms have small coefficients - the Rossby ($Ro = U/f_o \cdot L$) and Ekman [$E_H = A_H/(f_o \cdot L^2)$; $E_z = A_z/(f_o \cdot H^2)$] numbers, respectively, which were defined in Chapter 7.

Whenever higher order derivatives in the equations exist but have small coefficients we expect boundary layers to occur. These are relatively thin regions near the boundary where some of the terms with higher order derivatives will become important. Simple equations will hold in the interior of the fluid; in the oceanographic case they are the geostrophic equations. (The same approach is used in other fluid mechanics problems where rotation effects do not occur but we shall only look at the oceanographic case.) However, solutions to the geostrophic equations do not usually satisfy all the boundary conditions. Therefore near the boundaries some higher order term or terms must become large because the length scale normal to the boundary becomes small. Eventually some extra term or terms become large enough to allow the boundary conditions to be satisfied.

To use the boundary layer approach we re-scale the equations, changing the appropriate length scale to one relevant to the boundary layer thickness. The advantage of this approach is that only the higher order terms which matter will become important. Unimportant terms remain small and the 'boundary layer' equations will be simpler than the full equations and hopefully easier to solve. We expect the boundary layer effects to disappear a few boundary layer scale lengths from the boundary. We make the boundary layer solution satisfy the boundary conditions and then approach the interior solution as the boundary layer coordinate becomes very large. Because the boundary layer is very thin relative to the whole ocean and the interior solution is slowly varying, we can match to the interior solution effectively right at the boundary. Once one becomes convinced that this 'connecting' of boundary layer and interior regions is possible one can concentrate on examining a range of boundary layer solutions without worrying about the matching. This approach has been a very useful tool in analytical studies and has helped to improve our understanding of what dynamical effects may be important even when the solutions are too idealized to be applied in detail to the real ocean. The boundary layer approach may also show how the thickness of the boundary layer depends on the parameters of the system. Here we shall not give an exhaustive treatment of this approach but illustrate the method with a few fairly simple examples.

The Ekman solution for the wind-driven currents given earlier in the chapter is a boundary layer solution. The geostrophic equations describe the part of the motion associated with the horizontal pressure gradients. However, these solutions cannot satisfy the surface boundary condition (that the shear stress at the surface must be continuous) when there is an applied wind stress. Near the surface the vertical length scale for the vertical friction term becomes small and these vertical friction terms become large enough to balance the Coriolis term associated with the directly wind-driven flow. With the stress in the y-direction, as in the solution presented, the boundary condition at the surface is:

stress in the water at the surface in the y-direction
= wind stress at the surface in the y direction ,

i.e.,    $\rho \cdot A_z \cdot (\partial v/\partial z)_{z=0} = \tau_{y\eta}$ .

The reader may verify that the solution given satisfies this equation and that $(\partial u/\partial z)_{z=0} = 0$ since there is no stress in the x-direction. (In the more general case, $\rho \cdot A_z \cdot (\partial u/\partial z)_{z=0}$ = x component of the wind stress must hold also.) Indeed the speed of the surface flow $V_o$ in the solution is determined by this boundary condition.

The solution has the proper boundary layer character. In addition to satisfying the surface boundary condition, the speed of flow in the solution decays rapidly with depth and is essentially zero at the bottom of a very thin layer relative to the total ocean depth.

The matching with the interior is very straightforward in this case. If the Ekman flow is non-divergent then the Ekman and geostrophic flows are completely independent and may simply be added together as noted earlier. If the Ekman flow is convergent or divergent then the vertical velocity at the bottom of the Ekman Layer provides one of the boundary conditions which the geostrophic flow must satisfy as described in the section on the generalized Sverdrup equation earlier in the chapter. Because the Ekman layer is so thin one may, with negligible error, take this boundary condition on the geostrophic flow to be at z = 0 rather than at the 'bottom' of the Ekman layer.

## The Use of the Boundary Layer Approach to Obtain a Solution to Munk's Equation

Munk's equation (9.34) looks quite formidable. Although he did solve the complete equation for the rectangular ocean model, it is much easier to obtain an approximate, but sufficiently accurate, solution using the boundary layer approach. For the solution with the triangular ocean shape of Munk and Carrier the boundary layer approach had to be used because a solution to the full equation could not be found. Here we shall obtain the solution for the rectangular basin case. As noted earlier, in the interior $A \cdot \nabla^4 \psi$ can be neglected but in the side boundary-layers $\partial/\partial x$ terms (derivatives with respect to the co-ordinate normal to the boundary) become important so that boundary conditions of no flow through and no slip at the side boundaries can be satisfied.

Now instead of just looking at the sizes of terms we shall put Munk's equation (9.34) in non-dimensional form. This is a very useful procedure, often done in theoretical studies (both in geophysical fluid dynamics, i.e., oceanography and meteorology, and in other branches of fluid mechanics). The terms involving non-dimensional variables are then of order 1 and their non-dimensional coefficients, based on the scales of the system being studied, determine the importance of various terms. In the non-dimensional equations of motion the Rossby and Ekman numbers are important as mentioned earlier. Here we are going to non-dimensionalize Munk's form of the vertically integrated vorticity equation. We consider finding a solution applicable to a region such as the Gulf Stream gyre system. Now in the interior away from the western boundary (under Munk's assumptions given earlier) the simple Sverdrup balance (equation 9.21) holds. Integrating this equation with respect to x gives:

$$\psi_i \;=\; -\frac{1}{\beta} \int curl_z \underline{\tau}_{\eta} \cdot dx + C \tag{9.39}$$

where the subscript i indicates the stream function in the interior, C is a constant of integration and $curl_z \underline{\tau}_{\eta}$, as before, stands for $(\partial\tau_{y\eta}/\partial x - \partial\tau_{x\eta}/\partial y)$. Now the size of $\psi$ from equation 9.39 or 9.21, or more precisely the change of $\psi$ from the outer edge of the western boundary layer to the eastern side of the ocean, will be of order $\tau_o/\beta$, where $\tau_o$ is a typical wind-stress magnitude (or the total change in $\tau_{x\eta}$ from north to south in the gyre) and $\beta$, taken to be constant in this $\beta$-plane model, is the value of df/dy at the centre of the gyre. We note that the change in the value of $\psi$ across the western boundary must also be of the same size because the boundary current is returning the interior flow and the change in $\psi$ in each region gives a measure of the total transport. We put:

$$\psi \;=\; (\tau_o/\beta)\cdot \psi' \qquad \text{and} \qquad \underline{\tau}_{\eta} \;=\; \tau_o \cdot \underline{\tau}'_{\eta} \tag{9.40}$$

where the primes indicate non-dimensional variables with a range of order 1. The magnitude of $curl_z \underline{\tau}_{\eta}$ will be $\tau_o/L$ where L is the distance over which $\underline{\tau}_{\eta}$ changes by $\tau_o$, in this case the north-south dimension of the gyre. For x and y we put:

$$x \;=\; \xi \cdot W \qquad \text{and} \qquad y \;=\; \gamma \cdot L \tag{9.41}$$

where $\xi$ and $\gamma$ are non-dimensional variables. In the interior, W is O(L). (For simplicity we shall assume a square basin so x and y both go from 0 to L but the east-west width could be a multiple of L without changing the results.) In the western boundary, W << L because the boundary current is long and narrow. As x goes to a small fraction of L, $\xi$ will become very large if W is taken to be the width of the western boundary region. If we substitute Eqs. 9.40 and 9.41 into 9.34 we get:

$$\frac{A\cdot\tau_o}{\beta}\cdot\left(\frac{1}{W^4}\cdot\frac{\partial^4\psi'}{\partial\xi^4} + \frac{2}{W^2 L^2}\cdot\frac{\partial^4\psi'}{\partial\xi^2\partial\gamma^2} + \frac{1}{L^4}\cdot\frac{\partial^4\psi'}{\partial\gamma^4}\right) - \frac{\tau_o}{W}\cdot\frac{\partial\psi'}{\partial\xi}$$

$$= \tau_o\cdot\left(\frac{1}{W}\cdot\frac{\partial\tau'_{y\eta}}{\partial\xi} - \frac{1}{L}\cdot\frac{\partial\tau'_{x\eta}}{\partial\gamma}\right).$$

Multiplying by $W/\tau_o$ and some rearranging gives:

$$\frac{A}{\beta\cdot W^3}\cdot\left(\frac{\partial^4\psi'}{\partial\xi^4} + 2(\frac{W}{L})^2\cdot\frac{\partial^4\psi'}{\partial\xi^2\partial\gamma^2} + (\frac{W}{L})^4\cdot\frac{\partial^4\psi'}{\partial\gamma^4}\right) - \frac{\partial\psi'}{\partial\xi}$$

$$= \frac{W}{L}\cdot\left(\frac{L}{W}\cdot\frac{\partial\tau'_{y\eta}}{\partial\xi} - \frac{\partial\tau'_{x\eta}}{\partial\gamma}\right). \tag{9.42}$$

Note that $(L/W)\cdot(\partial\tau'_{y\eta}/\partial\xi)$ remains of O(1) both in the interior, where $W \approx L$, and in the boundary layer because the magnitude of $curl_z \underline{\tau}_{\eta}$ is the same in both regions. In the western boundary region where W << L and $\xi$ changes rapidly, $\partial\tau'_{y\eta}/\partial\xi$ becomes very small, of O(W/L). Recall that $\partial\tau_{x\eta}/\partial y$ is the dominant term in the wind-stress curl everywhere, so the term coming from $\partial \tau_{yn}/\partial x$ cannot become important no matter how we non-dimensionalize the equation.

Following Munk we take $A = 5 \times 10^3 \, m^2 \, s^{-1}$ and $\beta = 1.9 \times 10^{-11} \, m^{-1} \, s^{-1}$. For the interior take $W = 5 \times 10^6 \, m = 5,000 \, km$. Then $A/(\beta \cdot W^3) = 2 \times 10^{-6}$, so the friction terms are negligible and the simple Sverdrup balance holds. To make the friction terms of $O(1)$ to balance $\partial \psi'/\partial \xi$ we must take

$$W \;=\; (A/\beta)^{1/3} \;=\; (5 \times 10^3/1.9 \times 10^{-11})^{1/3} \approx 6 \times 10^4 \, m \;=\; 60 \, km \; .$$

This is an example of how the scaling can be used to find out how the 'width' of the western boundary current must depend on the other parameters in the system. Now with $W = 60 \, km$, the friction term is $O(1)$ and so is $\partial \psi'/\partial \xi$. However, the wind-stress term is now $O(W/L)$ or $O(0.01)$ and may be neglected to a good approximation. This result should not be too surprising. The wind distribution is quite uniform and symmetric; thus the local wind-driving in the western boundary region will be similar to that in the interior. With the much higher velocities and transport/unit width in the western boundary the relatively small local wind-induced transports can be ignored. Furthermore, not all of the higher order terms are equally important. With $W/L \approx 0.01$, only the first term is $O(1)$, the next largest one being $O(10^{-4})$ and therefore negligible. Thus in the western boundary we have, to a good approximation (about 1% which is rather good by geophysical fluid dynamics standards):

$$\frac{\partial^4 \psi'}{\partial \xi^4} \;-\; \frac{\partial \psi'}{\partial \xi} \;=\; 0 \; . \tag{9.43}$$

The use of the boundary layer approach has simplified Eq. 9.34 to Eq. 9.43, which shows the strength of the method.

Equation 9.43 has a quite simple solution (a sum of exponential functions which is usually the first type of solution tried for a linear differential equation with constant coefficients because it often works):

$$\psi' \;=\; C_o \;+\; \sum_{n=1}^{n=3} A_n \cdot \exp(a_n \cdot \xi) \tag{9.44}$$

where the $a_n$'s are the roots of $a_n^3 = 1$ which are $a_1 = 1$, $a_2 = -1/2 + j \cdot \sqrt{3}/2$ and $a_3 = -1/2 - j \cdot \sqrt{3}/2$ where $j = \sqrt{-1}$. Now $A_1 = 0$ must be chosen because $\exp \xi$ is divergent (becomes very large) as $\xi$ becomes large (mathematically as $\xi \to \infty$). The solution with $a_1 = 1$ can be used on the eastern boundary to satisfy the no-slip condition as we shall show presently. Using $\psi = 0$ (no flow through the boundary) and $\partial \psi/\partial x = 0$ ($M_y = 0$ or no slip or flow along the boundary) at $x = 0$, determines $A_2$ and $A_3$ and we have

$$\psi' \;=\; C_o \cdot \left[ 1 - \exp(-\xi/2) \cdot \left\{ \cos(\sqrt{3} \cdot \xi/2) + \frac{\sin(\sqrt{3} \cdot \xi/2)}{\sqrt{3}} \right\} \right] \tag{9.45}$$

$$=\; C_o \cdot T(\xi) \; .$$

$T(\xi)$ goes to 1 as $\xi$ becomes very large (as we go from the boundary layer to the interior).

Thus $\psi' \rightarrow C_o$ as we move to the interior and, in dimensional units, $C_o$ must be the value of the interior transport stream-function at the edge of the boundary layer.  Because $W/L \ll 1$, we can, to a good approximation, use $\psi_i$ at

$x = 0$.  Thus $\psi = \psi_i(x,y) \cdot T(x/W)$ and from before $\psi_i = -(1/\beta) \cdot \int_o^x \text{curl}_z \, \underline{\tau}_\eta \cdot dx + C$.

Using $\psi_i = 0$ at $x = L$, the eastern boundary, gives $C = \frac{1}{\beta} \int_o^L \text{curl}_z \underline{\tau}_\eta \cdot dx$ and finally:

$$\psi = \left[ (1/\beta) \cdot \int_x^L \text{curl}_z \, \underline{\tau}_\eta \cdot dx \right] \cdot T(x/W) \qquad (9.46)$$

where $T(x/W)$ is given by the expression multiplying $C_o$ in equation 9.45 with $\xi$ replaced by $x/W$.  To complete the solution fully, we add to the right-hand side of equation 9.46 $(W'/\beta) \cdot (\text{curl}_z \, \underline{\tau}_\eta)_{x=L} \cdot \exp\{(x-L)/W'\}$ which is negligible except near the eastern boundary and which makes $\partial\psi/\partial x$ vanish at $x = L$.  $W'$ is an eastern boundary width and $W' \ll L$.  Since A in the eastern boundary will probably be smaller than in the west, friction probably being weaker here $W'$ (proportional to $A^{1/3}$) is likely to be smaller than $W$ of the western boundary.

Now from the solution, the actual width of the western boundary current is three to four times $(A/\beta)^{1/3}$ or, for the values chosen by Munk, about 200 km.  This width is probably reasonable for the climatological average Gulf Stream or Kuroshio.  However, at any one time these streams are only 50-60 km wide.  Using the Munk theory with $A = (\beta \cdot W^3)$ suggests for the short-term average stream that A is of order $10^2 \, \text{m}^2 \, \text{s}^{-1}$.  Now the inertial or non-linear terms are of order $10^3/A = 10$ times the turbulent friction terms.  Thus for the short-term average western boundary current, inertial or non-linear effects are probably not negligible.

## A Simple Inertial Theory by Stommel

This is an idealized model used to see if a predominantly inertially controlled Gulf Stream is a reasonable approximation.  Stommel assumes a two-layer system.  The upper layer has density $\rho_1$ and is moving; the lower layer has density $\rho_2$ and is at rest.  The thickness of the upper layer, D, is 0 at the coast ($x = 0$) and increases to $D_o$ at the outer edge of the western boundary layer.  The x axis is taken across the stream and the y axis along it.

Before proceeding with Stommel's model we need to develop expressions for the pressure gradients in terms of gradients of the layer thickness.  Consider the pressure in the lower layer:

$$p = - \int_\eta^z \rho \cdot g \cdot dz = - \int_\eta^d \rho_1 \cdot g \cdot dz - \int_d^z \rho_2 \cdot g \cdot dz$$

where $\eta$ is the surface elevation from the rest state with $z = 0$ at the surface and d is the level of the interface between the layers measured from the $z = 0$ reference.  Now $\rho_1$ and $\rho_2$ are constants and may be taken outside the integrals giving:

$$p = \rho_1 \cdot g \cdot (\eta - d) + \rho_2 \cdot g \cdot (d - z)$$

$$= \rho_1 \cdot g \cdot \eta + (\rho_2 - \rho_1) \cdot g \cdot d - \rho_2 \cdot g \cdot z \quad .$$

Then $\frac{\partial p}{\partial x} = \rho_1 \cdot g \cdot \frac{\partial \eta}{\partial x} + (\rho_2 - \rho_1) \cdot g \cdot \frac{\partial d}{\partial x}$ in the lower layer. But in this
layer there is no flow (by assumption) and therefore the horizontal pressure
force must be zero, i.e.:

$$\frac{\partial d}{\partial x} = - \frac{\rho_1}{\rho_2 - \rho_1} \cdot \frac{\partial \eta}{\partial x} .$$

If we want to use the total upper layer thickness, taken to be positive,
$D = \eta - d$ and

$$\frac{\partial D}{\partial x} = \frac{\partial \eta}{\partial x} - \frac{\partial d}{\partial x} = \left(1 + \frac{\rho_1}{\rho_2 - \rho_1}\right) \cdot \frac{\partial \eta}{\partial x} = \frac{\rho_2}{\rho_2 - \rho_1} \cdot \frac{\partial \eta}{\partial x} .$$

Now $(\rho_2 - \rho_1) \ll \rho_1$, e.g., in the Gulf Stream $(\rho_2 - \rho_1) \approx 2 \times 10^{-3} \cdot \rho_1$. Thus the
slope of the interface is much larger than the slope of the surface and is of
opposite sign. Here we have derived the result for an idealized two-layer
system. In the more general case of continuous density variation, the results
would be similar. If the horizontal pressure gradients go to zero at depth,
the isopycnal slopes will mainly be opposite to the surface slope and much
larger.

In the upper layer above $z = d$:

$$p = \rho_1 \cdot g \cdot (\eta - z) \qquad \text{and} \qquad \frac{\partial p}{\partial x} = \rho_1 \cdot g \cdot \frac{\partial \eta}{\partial x} .$$

Also as we showed near the end of Chapter 8, the x momentum equation remains
geostrophic to a good approximation:

$$- f \cdot v = - \frac{1}{\rho_1} \cdot \frac{\partial p}{\partial x} = - \frac{\rho_2 - \rho_1}{\rho_2} \cdot g \cdot \frac{\partial D}{\partial x} = - g' \cdot \frac{\partial D}{\partial x} \qquad (9.47)$$

where $g' = g \cdot (\rho_2 - \rho_1)/\rho_2$ is termed the 'reduced gravity'. The total transport of the stream $T = \int_0^W Q_y \cdot dx$ where W is the value of x at the seaward
edge of the stream. As the upper layer is homogeneous, v is independent of
depth within the upper layer and zero below it, so $Q_y = v \cdot D$. Substituting
this expression and using equation 9.47 gives:

$$T = \int_0^W \frac{g'}{f} \cdot D \cdot \frac{\partial D}{\partial x} \cdot dx = \int_0^W \frac{g'}{f} \cdot \frac{\partial D^2/2}{\partial x} \cdot dx = \frac{g'}{f} \cdot \frac{D_o^2}{2}$$

since $D = 0$ at $x = 0$ and $D = D_o$ at $x = W$.

Now following Stommel we assume that the potential vorticity is essentially
constant. This assumption should be valid if friction effects are small
enough to be neglected. The important inertial terms are retained. Observations in the Gulf Stream do show that the potential vorticity remains nearly
constant. For the relative vorticity $\partial u/\partial y$ is negligible compared with $\partial v/\partial x$
both because $v \gg u$ and because the stream is long and narrow. Thus constant
potential vorticity (Eq. 9.38) reduces to $(f + \partial v/\partial x)/D = \text{constant} = f/D_o$ since
at the 'edge' of the stream $\partial v/\partial x$ will be small. Taking $\partial/\partial x$ of equation

9.47* and substituting for $\partial v/\partial x$ in the potential vorticity equation gives:

$$(f + \frac{g'}{f} \cdot \frac{\partial^2 D}{\partial x^2})/D = f/D_o .$$  Rearrangement gives  $\dfrac{\partial^2 D}{\partial x^2} = \dfrac{(D - D_o)}{\lambda^2}$

where $\lambda = \sqrt{(g' \cdot D_o)}/f$ is called the *Rossby radius of deformation* and gives a length scale based on the parameters of the system.  The solution is

$$D = D_o \cdot [1 - \exp(-x/\lambda)] , \quad v = \sqrt{g' \cdot D_o} \cdot \exp(-x/\lambda) .$$

With $D_o = 800\,m$, $f = 10^{-4}\,s^{-1}$, and $(\rho_2 - \rho_1)/\rho_2 \approx 2 \times 10^{-3}$, $T \approx 63\,Sv$ and the maximum value of $v = 4\,m\,s^{-1}$.  Also $\lambda = 40\,km$ which gives a length scale for the 'width' of the stream.

This simple inertial boundary layer model gives a transport closer to that of observations than the linear theory of Munk and, in the outer part of the stream, the velocity calculated from the model solution fits the velocity calculated using the geostrophic equation and observations of temperature and salinity reasonably well.  Near the inshore edge the observed velocity decreases while the model velocity continues to increase.

While the model is too simple to represent the actual Gulf Stream in detail it does indicate that inertial effects need to be included, particularly on the outer side of the stream south of where the transport is a maximum.  On the inshore edge, friction probably becomes important and beyond the latitude of maximum transport the stream shows much more meandering so a more complicated model is needed.  As mentioned earlier, Morgan and Charney developed more complete inertial models but these do not work north of the latitude of maximum transport either.

A frictionless, purely inertial, model is not likely to be completely satisfactory.  As originally noted by R.W. Stewart, the fact that the no-slip boundary condition is not satisfied allows the stream to transport relative vorticity.  Thus in the inertial models a considerable amount of relative vorticity is moved to the northwest corner of the gyre, probably making the dynamics used incorrect there and perhaps throughout much of the gyre.  Analytical models with both inertial and friction effects seem very difficult to deal with so, as noted before, recent attempts at complete ocean models have been done with numerical techniques which are discussed in Chapter 11.

The Rossby radius, $\lambda$, is named after Rossby who first introduced it in his wake stream theory of the Gulf Stream, an attempt to explain the countercurrents both inshore and offshore of the Gulf Stream.  Stommel's model indicates that it is an important length scale in the inertially dominated part of the main stream itself.  It also appears to be the relevant size scale for 'meso-scale eddies', the transient motions with quite large velocities which will be described briefly in Chapter 11.

---

\*  If the stream does not flow north-south so that x is not east-west, a term involving the variation of f with latitude would occur but it is negligible.

# CHAPTER 10
# Thermohaline Effects


THE DEEP CIRCULATION

This is much less well known and less well described dynamically than the
upper layer circulation.  From water property distributions it has been
accepted for some time that the main source of the deep water and its circulation is sinking off Greenland in the North Atlantic and/or in the Norwegian
Sea, and in the Weddell Sea area (South Atlantic).  The process is a thermohaline one, the sinking being due to increase of density consequent on decrease of temperature (North Atlantic) or on increase of salinity due to
freezing out of ice (South Atlantic).  For a long time the deep water motion
was envisaged as a very slow flow (millimetres per day) of the entire mass.
However, Wüst in 1955 showed for the Atlantic, both from the oxygen distribution and from geostrophic calculations, that much of the volume transport
occurred in a relatively narrow band on the west side of the ocean, to the
south in the North Atlantic Deep Water and to the north in the Antarctic
Bottom Water.  Other observations with Swallow floats and of ripple marks on
the deep ocean bottom in some areas add further evidence of fast currents in
the deep and bottom water, although the climatological average in most regions
is very slow.  So much is fact.

Stommel has advanced ideas for a model of the circulation of the deep waters.
He brings in another feature of the ocean structure - that the depth of the
thermocline at any locality remains substantially constant.  Because in low
latitudes there is a net annual inflow of heat through the surface into the
water, the upper warm layer, and its boundary the thermocline, should deepen
with time.  As this deepening does not happen, some mechanism must be opposing
the tendency, and Stommel suggests that this mechanism is slow upward flow of
cool deep water.  Continuity requires that the sinking water in the North and
South Atlantic must be balanced by rising, and Stommel suggests that while
the sinking is very localized, the rising is spread over most of the low and
middle latitude areas of the oceans.  His model of the character of the deep
circulation is shown in Fig. 10.1.

The sinking regions ($S_1$, $S_2$) are shown feeding relatively intense western
boundary currents (required by conservation of vorticity in a situation in
which the relative vorticity, $\zeta$, is known to be small in the interior).  Outward from these flow gentler geostrophic currents into the bodies of the
oceans to supply the slow upward flow to maintain the thermocline depth constant.  In the interior, upward motion causes D to increase; water moves poleward and the magnitude of f increases; $\zeta$ stays small.  To get back south or
north as necessary with $\zeta$ small requires input of vorticity of the appropriate
sign.  This input may be achieved with a strong flow and shear on the west
again.  Fig. 10.2 shows that the strong flow and shear must be on the west

134

Fig. 10.1.    Model for deep ocean circulation after Stommel
              (*Deep-Sea Res.*, 5, 82, 1958).


rather than on the east when the boundary return flow is to the south, since
southward flow requires input of negative vorticity to keep $\zeta$ small.

Fig. 10.2.    Relative vorticity supplied by velocity shear at west and east
              sides of the ocean.

In discussing the western boundary surface currents in the last chapter we showed that a northward return flow also had to be on the west. Observations reported in 1977 by Warren indicate an additional feature of the deep flow in the Indian Ocean. Earlier observations had shown the northward flow expected from Stommel's model along the west side of the ocean off Madagascar while the new observations showed a northward flow along the Ninetyeast Ridge which rises to about 4,000 m depth along that meridian which is to the east of the centre of the Indian Ocean. The inference is that a western boundary type flow can be associated with the mid-ocean ridges, if high enough above the bottom, as well as with the western boundary itself.

An aspect of this model relates to the different strengths of the surface layer western boundary currents. Where the deep currents are opposite in direction to the surface flows, the latter may be expected to be stronger to maintain continuity. The sinking water must be replaced by water which has come up through the thermocline and returns in the upper layer. Conservation of vorticity again requires western boundary currents in the return flow. Above the thermocline the upward flow causes D to decrease. To keep $\zeta$ small the flow is equatorward, i.e., opposite to the deep flow and likewise to conserve vorticity the western boundary flows in the upper layer will be opposite to those in the lower layer. The strong Gulf Stream to the northeast in the upper layer is consistent with the strong southwest flow in the deep water from $S_1$ (Fig. 10.1), while the less strong Kuroshio is associated with the weaker deep flow in the Pacific, although the thermohaline flow does still enhance the Kuroshio in the surface region compared with the purely wind-driven values. (Note that there is no large source of deep water in the Pacific. There is now believed to be some outflow from the Ross Sea in the Antarctic but the volume appears to be much smaller than that from the Weddell Sea.) Again, the relatively weak southward Brazil Current in the upper layer in the South Atlantic is consistent with the southward deep flow below it. Stommel suggests that the flow in the deep current under the Gulf Stream is about 30 Sv. The equal surface return flow would then almost double the Gulf Stream transport associated with wind driving. Even addition of such a thermohaline flow to the wind-driven flow does not make the transport large enough to match values from recent fairly direct measurements.

Stommel makes it clear that he does not regard the above as a *theory* but as a *model* for quantitative study which might form the basis for a theory. A limited number of deep current measurements with Swallow floats have in some cases supported the model and in some cases opposed it. However, the number of measurements yet available is small and must be much increased before deciding whether it would be profitable to develop a more detailed theory from the model or to develop a new model and theory.

To produce a theory with thermohaline effects included may require taking T and S (and hence $\rho$) not as given by observations but as unknowns which must be solved for, in the problem, along with velocity. If T and S are to be unknowns then we require equations for them.

## EQUATIONS FOR SALT AND TEMPERATURE (HEAT) CONSERVATION

The differential equations for salinity and temperature (representing heat) are:

$$\frac{dS}{dt} = \kappa_S \cdot \nabla^2 S \qquad\qquad\qquad (10.1)$$

$$\frac{dT}{dt} = \kappa_T \cdot \nabla^2 T + Q_T \qquad\qquad (10.2)$$

where $\kappa_S$ and $\kappa_T$ are molecular kinematic *diffusivities* for salt and for temperature (or heat) respectively, and $\nabla^2 = \partial^2/\partial x^2 + \partial^2/\partial y^2 + \partial^2/\partial z^2$. The diffusivities have the same units as kinematic viscosity ($m^2 s^{-1}$). $\kappa_T$ is about $\nu/10$ and $\kappa_S$ is about $\nu/1000$*.

It has been assumed that $\kappa_S$ and $\kappa_T$ vary with position slowly enough that such variations can be ignored, i.e., terms such as $\partial[\kappa_S \cdot (\partial S/\partial x)]/\partial x$ which arise in deriving these equations are approximated by $\kappa_S \cdot (\partial^2 S/\partial x^2)$. $Q_T$ represents a source function. For example solar radiation is absorbed over a significant depth range and causes heating and $Q_T$ represents such effects. A similar function for S is not required because processes affecting salinity occur only at boundaries, e.g., river inputs, effects of freezing or the difference between evaporation and precipitation. Such effects would be included as boundary conditions, i.e., for the river one would specify the incoming velocity, salinity (if any) and temperature of the river water, for the other two processes one would specify the equivalent salt flux at the boundary as mass of salt per unit area per unit time. $Q_T$ and boundary conditions will be taken to be given. (In practice, detailed specification might be difficult. If we were considering vertical averages between a number of levels, as in a finite difference numerical model, and the solar radiation were all absorbed in the top layer, then the solar radiation could be included as part of the net heat flux at the surface.)

Note that on the left-hand side we have the total derivative. These equations apply to an individual, small (mathematically infinitesimal) fluid element. Indeed they may be easily derived by considering the net flux of the quantity of interest in all directions for a small element and equating it with the total rate of change of the quantity inside the element. For example, the flux of salt in the x-direction is $-\kappa_S \cdot (\partial S/\partial x)$ and the net flux in (as $\delta x \to 0$) is $\partial(\kappa_S \cdot \partial S/\partial x)/\partial x$ and similarly for the other directions. These equations are similar to the Navier-Stokes equations for momentum but are rather simpler in that many of the types of terms occurring in the latter do not appear (e.g., terms like those involving pressure, Coriolis force and gravity).

Note also that a fluid element can only change its salinity by molecular processes. Except where $Q_T$ is important, the same is true for temperature (and thus heat). Molecular viscous effects are usually important for fluid elements too, although pressure, Coriolis and gravitational forces are also acting. When we deal with the averaged equations, as we are forced to do because we

---

* These differences arise from the molecular nature of a liquid (water in this case). It is harder for the molecules to exchange kinetic energy (which determines the temperature) than momentum (for frictional stress). It is even harder to move a different type of molecule (e.g., salt) through the fairly closely packed water molecules. For gases at standard temperature and pressure, which are loosely packed, the kinematic viscosity and the diffusivities are approximately equal.

cannot solve the equations for the instantaneous quantities, such details get
obscured.  Nevertheless, it remains true that molecular effects are always
important and often dominant for fluid elements.  When these elements are in
turbulent flow, the turbulence, in 'stirring' the fluid, makes the instantaneous property gradients very large and greatly increases the rate of change
of properties of the elements compared with the rate in a non-turbulent fluid
with a comparable mean gradient.  In the averaged equations the mixing is
described by the Reynolds stresses for momentum (and Reynolds fluxes for salt
and temperature as we shall show presently).  To get actual solutions these
Reynolds stresses and fluxes are related to the mean gradients by introducing
eddy viscosity (or diffusivity for salt and temperature) as a necessary
approximation for rather detailed processes which we cannot deal with
completely.

Equations like 10.1 or 10.2 also apply to other scalar properties.  For example, oxygen concentration would be represented by an equation of the form of
10.2 with oxygen concentration replacing T.  $Q_{ox}$ could be either a source or
sink depending on what biological processes are occurring.

When we regarded S and T as given, we had four equations in four unknowns
(Chapter 6).  We have now added equations for S and T.  With S and T unknown,
$\rho$ (or $\alpha$) must also be regarded as unknown but it can be obtained from S, T
and p through the equation of state.  (While this relation is often given in
table form it is possible to write it as a polynomial in S, T and p for
numerical calculation if needed.)  Thus we now have seven equations in seven
unknowns.  With appropriate boundary conditions (which may include initial
conditions if necessary) it is possible in principle to obtain solutions.  In
practice, the non-linear terms in the Navier-Stokes equations present difficulties as we have already seen.  The addition of three more equations and
unknowns is not likely to help, and indeed things get worse as we shall see.


## EQUATIONS FOR THE AVERAGE SALINITY AND TEMPERATURE

Equations 10.1 and 10.2 apply to the total instantaneous values of S and T.
If we recall the Eulerian form of the total derivative (e.g., $dS/dt = \partial S/\partial t +
u \cdot \partial S/\partial x + v \cdot \partial S/\partial y + w \cdot \partial S/\partial z$) the problem is clear.  We must use Eulerian
equations for velocity and hence for S and T also.  Terms such as $u \cdot \partial S/\partial x$
arise; they are called advective terms because they represent changes caused
by the motion of the fluid.  Here u, v and w are the total instantaneous
velocity components.  We cannot solve for them in practice as discussed in
Chapter 7; therefore we cannot use the equations 10.1 and 10.2 directly.  (As
with velocity, the fact that boundary and initial conditions for S and T would
never be known sufficiently well to solve for instantaneous values is a further difficulty.)  The problem arises from the presence of the advective terms
involving cross-products of velocity components and the scalar quantity together with the inability to calculate the velocity.

As we did in Chapter 7, we adopt Reynolds' approach of splitting the total
quantities into mean and fluctuating parts:  $S = \bar{S} + S'$, $u = \bar{u} + u'$, etc.,
and take the average of the equation.*  Recalling that averages of terms

---

*The procedure here is very similar to that used in Chapter 7 to obtain the
Reynolds equation, so less detail will be given here.

involving a single fluctuating quantity vanish, we get from equation 10.1:

$$\frac{\partial \bar{S}}{\partial t} + \bar{u} \cdot \frac{\partial \bar{S}}{\partial x} + \bar{v} \cdot \frac{\partial \bar{S}}{\partial y} + \bar{w} \cdot \frac{\partial \bar{S}}{\partial z} + \overline{u' \cdot \frac{\partial S'}{\partial x}} + \overline{v' \cdot \frac{\partial S'}{\partial y}} + \overline{w' \cdot \frac{\partial S'}{\partial z}}$$

$$= \kappa_S \cdot \nabla^2 \bar{S} \; . \qquad (10.3)$$

The first four terms may be written as $d\bar{S}/dt$ for the total derivative follow-ing the *mean* flow. The next three represent the effects of turbulence on the salinity field. The diffusivity term looks the same as before except that $\bar{S}$ replaces S.

## Reynolds Fluxes and Eddy Diffusivity

Using the continuity equation for the fluctuating velocity ($\nabla \cdot \underline{V}' = 0$) we can rewrite the turbulent terms by adding $S' \cdot \nabla \cdot \underline{V}'$ (which = 0) to the three terms in 10.3 which become $\partial(\overline{u'S'})/\partial x + \partial(\overline{v'S'})/\partial y + \partial(\overline{w'S'})/\partial z$. By analogy with the molecular case we suppose that the turbulent fluxes $\overline{u'S'}$, $\overline{v'S'}$, $\overline{w'S'}$ (also called Reynolds fluxes because they arise when the Reynolds approach is used) are related to the mean gradients in a similar fashion. (As in Chapter 7 we use the simplest analogy, leaving more complex formula-tions to more advanced texts.) The analogy gives:

$$\overline{u'S'} = -K_{Sx} \cdot \partial \bar{S}/\partial x; \quad \overline{v'S'} = -K_{Sy} \cdot \partial \bar{S}/\partial y; \quad \overline{w'S'} = -K_{Sz} \cdot \partial \bar{S}/\partial z$$

$$(10.4)$$

where $K_{Sx}$, $K_{Sy}$ and $K_{Sz}$ are kinematic eddy diffusivities (units $m^2 s^{-1}$).

Now the turbulent mixing is dominated by the turbulent flow field so it is common to assume that the eddy diffusivity is the same for all scalars (un-like the molecular values). Because of the static stability, $K_{Sz}$ will be much smaller than $K_{Sx}$, $K_{Sy}$ but these two should be similar. Thus we replace $K_{Sz}$ by $K_z$, vertical eddy diffusivity, and the other two by $K_H$, the horizontal eddy diffusivity. The ranges of values for $K_z$ and $K_H$ are similar to those for $A_z$ and $A_H$ respectively (the eddy viscosities) because they are properties of the turbulent flow field. In a particular case, $K_z$ and $A_z$ or $K_H$ and $A_H$ may not have the same values but they probably have the same order of magni-tude. Finally, neglecting the variations of the K's with space coordinates, neglecting the $\kappa_S$ term compared with the turbulence terms and dropping the overbar for simplicity, equation 10.3 becomes:

$$\frac{dS}{dt} = K_H \cdot \left(\frac{\partial^2 S}{\partial x^2} + \frac{\partial^2 S}{\partial y^2}\right) + K_z \cdot \frac{\partial^2 S}{\partial z^2} \; . \qquad (10.5)$$

Here S is now the average salinity. Equation 10.5 looks quite similar to 10.1 for the instantaneous salinity.

In the same manner, an equation for the average temperature may be obtained. Overbars are omitted but all quantities are now averages including $Q_T$:

$$\frac{dT}{dt} = K_H \cdot \left(\frac{\partial^2 T}{\partial x^2} + \frac{\partial^2 T}{\partial y^2}\right) + K_z \cdot \frac{\partial^2 T}{\partial z^2} + Q_T \; . \qquad (10.6)$$

However, we have made bold assumptions in inserting the eddy diffusivities. All the cautions in Chapter 7 with regard to the eddy viscosities apply. Any

results which depend strongly on the assumption of the eddy diffusivity (or
viscosity) representation must be viewed with suspicion until confirmed by
observations.  Use of $K_z = A_z$ in the atmospheric surface layer is an example
of a flow for which the approach does work well.

## THERMOCLINES AND THE THERMOHALINE CIRCULATION

Let us consider the steady state case and ignore $Q_T$ because we are interested
in the main thermocline.  Then equation 10.6 becomes:

$$\underline{V}_H \cdot \nabla_H T + w \cdot \partial T/\partial z = K_H \cdot \nabla_H^2 T + K_z \cdot \partial^2 T/\partial z^2 \qquad (10.7)$$

where $\underline{V}_H$ is the horizontal velocity ($\underline{i} \cdot u + \underline{j} \cdot v$) and $\nabla_H$ and $\nabla_H^2$ are the horizontal gradient and Laplacian respectively.  In words, equation 10.7 states that
horizontal advection plus vertical advection equals horizontal plus vertical
diffusion.

Our knowledge of the deep circulation is not sufficient to allow us to drop
any of the terms as being small.  Stommel's conceptual model suggests that
both advective terms are needed in a thermohaline circulation theory and at
least the vertical diffusion term.  Lateral diffusion may well be important
too.  It is possible to try to balance pairs of these four terms while neglecting the other two to see what solutions are possible.  At least five of
the six possibilities have been tried and more than one of the possible
balances can produce a reasonable looking thermocline structure.

The idea that vertical advection is balanced mainly by vertical diffusion with
the other terms being fairly small has been considered a reasonable possibility for a long time.  Assuming that this balance is correct we get:

$$\frac{\partial^2 T}{\partial z^2} = \frac{w}{K_z} \cdot \frac{\partial T}{\partial z} \cdot \qquad (10.8)$$

If we assume $w/K_z$ to be independent of z, then the solution is $T = T_o \cdot \exp$
$(wz/K_z)$ where $T_o$ is the temperature at $z = 0$ which is taken to be at the
bottom of the mixed layer.  Adding a mixed layer on top and adjusting $w/K_z$
one can produce a reasonable fit to observed vertical temperature profiles.

In the interior of the ocean we can use the geostrophic approximation
(equation 8.9).  If we cross-differentiate to eliminate the pressure terms
(i.e., take $\partial/\partial y$ of the x equation and $-\partial/\partial x$ of the y equation) we get:

$$\beta \cdot v + f \cdot (\partial u/\partial x + \partial v/\partial y) = 0 . \quad \text{Using the continuity equation gives:}$$

$$\beta \cdot v = f \cdot (\partial w/\partial z) .$$

To have north-south flow, w must vary with z.  As argued earlier, we must have
north-south flow in the interior to keep the relative vorticity small.  In
the interior, w is thought to be upward.  It could increase from zero at great
depth to a maximum in the thermocline and decrease to zero at the base of the
mixed layer (or to $w_E$ if we include wind-driving).  This vertical dependence
of w would give the north-south flows needed to keep relative vorticity small.
For our solution to equation 10.8 we require $w/K_z$ to be constant.  Thus $K_z$
would have to be a maximum in the thermocline too.  This behaviour for $K_z$
seems contrary to the expectation that $K_z$ will be lower in the region of

strongest static stability.  However, we do not know how $K_z$ depends on height.
Also there are likely to be internal waves in the thermocline and breaking of
these waves could lead to sufficient mixing to make $K_z$ larger there.

To keep things in perspective, remember that reasonable looking solutions have
been found with $K_z = 0$.  Clearly it is difficult to find satisfactory solutions
even in the interior.  To obtain a closed basin solution (which has not yet
been achieved) lateral boundary layers would have to be added.

Further discussion of thermocline and thermohaline circulation theories is
left for more advanced texts, both because the mathematics rapidly becomes
fairly complicated and because the observational data are so limited that it
is difficult to tell whether or not the theories are realistic.  A review
article by P. Welander provides a starting point in the literature (Phil.
Trans., Roy. Soc. Lond., Ser. A, 270,69-73, 1971).  Numerical modelling may
eventually help in obtaining more complete analytical solutions by showing
which terms must be retained and which may be neglected in the equations.

## THE MIXED LAYER OF THE OCEAN

The top few tens of metres of the ocean are usually observed to be fairly well
mixed, i.e., the temperature and salinity are fairly uniform.  Below this
region there is a thermocline (and perhaps a halocline) and hence a pycnocline region.  The top layer is the oceanic planetary boundary layer where
vertical friction effects are important.  It is also called, as we did earlier,
the Ekman layer.  The dynamics governing the formation of this layer are of
considerable interest.  Convergences and divergences in the layer lead to
circulations in the deeper water (Ekman pumping effect).  This is also the
region of (biological) primary productivity.  The depth of the layer and
mixing up of nutrients from below will be important factors in determining
the productivity.

In addition, there are meteorological effects both for weather and climate.
Much of the solar radiation, which is the ultimate source of energy for both
atmospheric and oceanic motions (except tides), is first absorbed in the
ocean's upper layer.  A large part of the atmosphere's energy supply comes
from heat exchange with this layer, mainly in the form of the latent heat
of the water evaporated at the surface which is released when the water con-
denses higher in the atmosphere.  For weather forecasting for a day or two
ahead these energy inputs can be ignored.  As the forecast period is increased
the energy inputs become increasingly important.  For an atmospheric model,
the bottom boundary conditions require knowledge of the surface temperature
(and the albedo which is reasonably well known for the ocean).  From the
surface temperature and humidity (which is determined by surface temperature) and the predicted air temperature, humidity and wind at the lowest level
in the atmospheric model one can calculate surface friction, heat fluxes and
back radiation although there is still some uncertainty in the coefficients
used (e.g., the drag coefficient and equivalent coefficients for heat fluxes).
To get the surface temperature, the evolution of the ocean mixed layer must
be predicted too.  For weather, or relatively short time scales of a few
days to perhaps a month, vertical transfers probably dominate and horizontal
advection and diffusion may be neglected.  For climatic time-scales of a
month to a few years the whole upper layer circulation (based on present
mean values) would need to be included because a sizable part of the poleward

heat transport required to maintain the global heat balance is provided by the
ocean circulation in the top few hundred metres. For decades to centuries the
whole ocean circulation must be included - a formidable task as we have shown.

Here we shall consider the short time-scale problem only. Without going into
the mathematical details we shall try to indicate the processes which are included and the approach being taken in this continuing aspect of dynamic
oceanography research. The conceptual model presented here comes from P.
Niiler (J. Mar. Res., 33, 405-422, 1975). The temperature is assumed uniform
within the mixed layer with a thin transition zone at the bottom where the
temperature changes rapidly to the value in the thermocline below the mixed
layer. (If salinity variations are important in determining density the effect
may be included by defining an equivalent temperature.) The temperature
structure below the mixed layer must be specified (presumably it can be
obtained from observations with geographical and seasonal variations included)
but hopefully it will be slowly varying on the time scale for which predictions are required. The velocity is also assumed to be independent of depth
throughout the bulk of the layer. There is a jump in the transition zone at
the bottom to the value of zero in the non-turbulent thermocline region. (If
a geostrophic flow is present in this region, it would have to be specified
and the velocity considered would be the difference from the geostrophic
value. There is also a thin shear zone in the surface wave zone. The stress
at the surface is specified.

With horizontal gradients assumed negligible, by continuity and w = 0 at the
surface there is no vertical velocity and all the non-linear terms involving
the mean velocity vanish. The equations for the horizontal velocity are:

$$\frac{\partial u}{\partial t} - f \cdot v = - \frac{\partial}{\partial z} \overline{u'w'} + F_x$$

$$\frac{\partial v}{\partial t} + f \cdot u = - \frac{\partial}{\partial z} \overline{v'w'} + F_y \ .$$

$$(10.10)$$

Primed quantities are fluctuations, unprimed are means. $F_x$, $F_y$ are damping
terms added to make inertial oscillations die out.

In the temperature equation the advective terms vanish because of the assumption of no horizontal gradients and the consequence that w = 0. The source
term (solar radiation) is included in the specified surface heat flux since
we are treating the layer as a whole and the temperature equation is:

$$\frac{\partial T}{\partial t} + \frac{\partial}{\partial z} \overline{w'T'} = 0 \ .$$

$$(10.11)$$

Now in the layer u, v, T, $F_x$ and $F_y$ are independent of z (by assumption).
Thus $\partial/\partial z$ of $\overline{u'w'}$, $\overline{v'w'}$ and $\overline{w'T'}$ must also be independent of z and the stresses
and the heat flux are linear functions of z, going from the specified values
at the surface to the values at the top of the transition zone required to
bring fluid being mixed into the layer to the values of u, v and T within the
layer. These expressions for $\overline{u'w'}$, $\overline{v'w'}$ and $\overline{w'T'}$ may be put into equations
10.10 and 10.11. In this process an additional unknown, the layer depth h,
is introduced. Thus we have three equations in four unknowns and therefore
a closure problem as is usual in a turbulence problem. The equation for
conservation of turbulent kinetic energy is added and to close the system it

is *assumed* that a fraction of the energy input by the wind coming from the
upper shear layer is diffused downward and used to mix fluid up from below
the mixed layer, the remainder being dissipated.  If the surface is being
cooled then this convective instability can provide mixing energy; if it is
being heated some of the mechanical mixing energy is used to overcome this
stabilizing effect.  The velocity 'jump' at the bottom may also cause mixing.
It causes layer deepening only if the layer is initially sufficiently thin
and only for the first half-pendulum-day after a change in the wind.  By that
time the initial rapid deepening has thickened the layer and reduced the
velocity jump; the Richardson number (based on the velocity and temperature
changes across the transition zone) becomes too large and the static stability
prevents mixing from this source.  If the initial layer depth is too large
(greater than the depth to which it would have mixed if it were initially
thin) the velocity 'jump' at the bottom is never large enough to produce
mixing.  The fraction of turbulent energy from the surface zone which is
assumed to diffuse down and cause mixing causes a slower continual deepening
of the layer after the initial rapid deepening (if it occurs).

This model and similar models have not been tested very much with real observations because data sufficiently detailed to test them fully are not available in the historical records.  What testing has been done suggests that
the approach has promise.  It seems to work fairly well during the heating
season when the summer or seasonal thermocline is being formed.  During the
cooling season the continual slow erosion from the constant fraction of surface generated turbulent energy, which is assumed to get to the bottom no
matter how deep, makes the model mixed-layer too deep and the annual cycle
does not close.  In reality it is probable that when the layer gets sufficiently deep the surface generated turbulent energy is all dissipated within
the layer and none is left to deepen the layer further.

Attempts to modify the closure scheme and to examine the possible importance
of the diurnal cycle in the heating have been made in a recent model which
produces a closed annual cycle (for idealized rather than real forcing) and
shows that added diurnal variation does affect the results.  Special observational programmes to obtain better field data are in progress.  Eventually,
with further testing and modification a satisfactory model should emerge.
By 'satisfactory' we mean that it predicts what we want to know with acceptable accuracy - it need not reproduce all the details of the field data.

# CHAPTER 11
# Numerical Models

## INTRODUCTION

Models such as those of Stommel and Munk described in Chapter 9 were attempts
to reproduce the general or climatological (long-term average) circulation and
seemed to give qualitatively correct results. However, the actual magnitude
of the transport seemed too small. Stommel, in "The Gulf Stream", suggested
that addition of thermohaline effects and perhaps some adjustment of the wind
stress, because of its uncertainty, might give better agreement between Munk's
model and the observations. At that time the transport estimates were based
on geostrophic calculations and so could be changed by varying the level of no
motion which could also help to reduce the discrepancy. Subsequently, more
direct transport measurements have shown that the discrepancy is even bigger
than had been thought and probably cannot be explained by simple addition of
thermohaline circulation and adjustment of the wind stress. Of course, it
had been known for a long time that if one looks at the Gulf Stream and similar currents over a short period of time they are much narrower and faster
than the currents in the models. Nowadays, one can locate such features from
surface temperature measurements obtained by satellites because the surface
temperature field gives an indication of the location of the current. The
currents are also known to meander, that is their position changes with time,
and they follow quite a curved path at times. It was hoped that if these
time-varying effects were averaged out then this average current would look
like the currents in the models.

Stommel also suggested another possible cause for the discrepancy which needed
further examination. As noted, the actual 'instantaneous' Gulf Stream is
quite narrow and strong. One can look at the sizes of the terms in the
equations of motion just as we did in Chapter 7 for the interior flow and as
we did at the end of Chapter 8 for a strong current such as the Gulf Stream.
We took the y coordinate along the stream and the x coordinate across the
stream. When we examined the x momentum equation, the Coriolis term $f \cdot v$
still dominated because v is so large; it must be balanced by the pressure
gradient term $\alpha \cdot (\partial p / \partial x)$ because none of the other terms are large enough.
Thus, the geostrophic balance still holds for this equation and this fact is
the basis for using the geostrophic calculation to obtain the v or downstream
velocity component as we noted before. In the y momentum equation we expect
friction to be important and estimate the eddy viscosity from the observed
width, W, of the stream using the relationship from Munk's model $A_H = \beta \cdot W^3$
developed at the end of Chapter 9. If we examine the other terms in this
equation using this estimate for $A_H$, rather than the maximum value as we did
in Chapter 8, the non-linear terms are larger than the friction terms as noted
in Chapter 9 and both effects are important (as the reader may easily verify
for himself). Thus to model the Gulf Stream as it actually occurs, non-

144

linear terms must be included.  Furthermore, if including these terms makes
the stream stronger than it would be without them, a long-time average, while
similar to the Munk model, would have a larger transport.  Some attempts were
made to actually solve the equations analytically with the non-linear terms
included as noted in Chapter 9.  However, only solutions restricted to very
special cases could be obtained.  It was the desire to include the non-linear
terms in a more adequate fashion that led to the early numerical models.  Thus
one tried to obtain solutions using approximate equations which could be
solved numerically using a computer; hence the term *numerical modelling*.

There were other parallel developments which also led to numerical modelling
attempts.  Numerical or computer modelling of the atmospheric circulation had
begun before these attempts at numerical models of the ocean were made.  Nowadays such models of the atmosphere are used to aid in operational weather
forecasting as well as for research purposes to try to understand the atmospheric circulation better.  Quite powerful computers are required for this
purpose.  The usefulness of such models for atmospheric circulation naturally
led to the idea of using them to model the ocean circulation.  Because such
models, particularly the more detailed ones, require very powerful and fast
computers the ocean modelling is often done in the same laboratories that are
doing the atmospheric modelling.  The atmospheric modellers are also interested in the ocean and in modelling it because the ocean forms 70% of the
bottom boundary of the atmospheric models.

Another parallel development is numerical modelling of coastal regions, semi-
enclosed seas and estuaries.  This type of modelling has developed as an
extension of numerical modelling of rivers.  It is often done for engineering
purposes as opposed to research purposes.  Examples are for predicting the
effects of construction  or determining the best location for waste disposal.
Prediction of tidal elevations and currents as an aid to navigation is
another example.  In addition, prediction of storm surges produced by the
combined effects of wind and tide is also of considerable practical interest.

Since this introduction to dynamical oceanography is primarily concerned with
the main features of the ocean circulation, neither coastal nor atmospheric
modelling will be described, but the results of these types of models may be
mentioned when they help us to understand the ocean models or when the results
of such modelling provide some insight to the future potential of ocean
models.

Two broad categories of models of the ocean circulation may be defined:  *mechanistic* models and *simulation* models.  In mechanistic models the geometry is
made as simple as possible and no terms are included in the model equations
which are not essential to answering the question at hand.  In such models
one may try to examine the importance, for example, of the non-linear terms
or the effect of bottom topography on the circulation.  In simulation models
one attempts to reproduce the circulation of an actual ocean for comparison
with oceanographic observations.  In such models the actual geometry of the
ocean basins is included with all the possible driving effects and terms in
the equations.  Simulation models produce an enormous amount of data and it
may be difficult to untangle which dynamic effects are most important, and
a great deal of analysis of the output of the model is required.  Both types
of models are important in increasing our understanding of ocean circulation.
Mechanistic models are easier to interpret and help to advance our understanding of particular aspects of ocean circulation dynamics.  On the other
hand, simulation models allow direct comparison with nature which is essential

to prove that ocean models really do represent nature and that the mechanistic models which form the basis of the simulation models lead to an understanding of the dynamics of the real world.

A detailed discussion of numerical methods is of course beyond the scope of this book. However, because many of the readers may be unfamiliar with the approaches taken, the next section will give a very general description of the more common approaches used. Then we will describe a number of ocean models and their results.

### NUMERICAL METHODS

The method most commonly used in numerical models, particularly in the simulation models, is that of *finite differences*. We have already made use of some finite difference approximations, for example, in calculating the horizontal divergence approximately in Chapter 4.

In such methods we do not attempt to find equations which will allow us to give the velocity at a particular time at *any* position, but to find values on a grid of points. For example, a two-dimensional grid may be used if we are considering the case in which we either deal with variables vertically averaged over the water column or assume that there is no vertical variation in the flow, or assume that there is no variation in the cross-stream direction and consider only variations in the along-stream and vertical directions. We need a three-dimensional grid if we allow variations in all spatial dimensions. We shall also calculate values at discrete times $t_1$, $t_2$ · · · $t_n$, rather than continuously.

To calculate values at a new time from values at a previous time, using the equations of motion, we have to be able to calculate spatial gradients of quantities. For example, suppose we want to determine $\partial u/\partial x$ at the point $x_j$. We could approximate this value by $(u_j - u_{j-1})/\Delta x$, where $\Delta x$ is the grid spacing in the x direction, $u_j$ is the value at $x_j$ and $u_{j-1}$ is the value at $x_{j-1}$. This is called a *backward difference*. We could use $(u_{j+1} - u_j)/\Delta x$, a *forward difference*. A Taylor series expansion around $x_j$ gives:

$$u_{j+1} - u_j = \frac{\partial u}{\partial x} \cdot \Delta x + \frac{1}{2} \cdot \frac{\partial^2 u}{\partial x^2} \cdot (\Delta x)^2 + \frac{1}{6} \cdot \frac{\partial^3 u}{\partial x^3} \cdot (\Delta x)^3 + \frac{1}{24} \cdot \frac{\partial^4 u}{\partial x^4} \cdot (\Delta x)^4 + 0(\Delta x)^5,$$

$$u_j - u_{j-1} = \frac{\partial u}{\partial x} \cdot \Delta x - \frac{1}{2} \cdot \frac{\partial^2 u}{\partial x^2} \cdot (\Delta x)^2 + \frac{1}{6} \cdot \frac{\partial^3 u}{\partial x^3} \cdot (\Delta x)^3 - \frac{1}{24} \cdot \frac{\partial^4 u}{\partial x^4} \cdot (\Delta x)^4 + 0(\Delta x)^5,$$

where derivatives are evaluated at $x_j$, and $0(a)^n$ means that the term is a finite number times $a^n$, called 'of order $a^n$'. The largest term of the error in the gradient in either the forward or backward approximations of $\partial u/\partial x$ is $(\partial^2 u/\partial x^2) \cdot \Delta x/2$, that is proportional to $\Delta x$. Such schemes are said to be first order accurate, because the error in the approximation depends on $\Delta x$ to the first power. A more accurate approximation to $(\partial u/\partial x)$, obtained by adding the equations and dividing by $2\Delta x$, is:

$$(u_{j+1} - u_{j-1})/2 \cdot \Delta x = \frac{\partial u}{\partial x} + \frac{1}{6} \cdot \frac{\partial^3 u}{\partial x^3} \cdot (\Delta x)^2 + 0(\Delta x)^4.$$

The largest error term is now $0(\Delta x)^2$ and this scheme, also called a *centred difference* since it uses a value on either side of $x_j$, is called second order

accurate. The next higher term is $O(\Delta x)^4$ because the terms $O(\Delta x)^3$ cancel.
Higher accuracy approximations can be obtained using two points on either
side of the point where we want an approximation to the derivative. Again
using Taylor series expansions we have:

$$(u_{j+2} - u_{j-2})/4 \cdot \Delta x = \frac{\partial u}{\partial x} + \frac{4}{6} \cdot \frac{\partial^3 u}{\partial x^3} \cdot (\Delta x)^2 + O(\Delta x)^4 \ .$$

Then $\frac{\partial u}{\partial x} = \frac{4}{3} \cdot \left( \frac{u_{j+1} - u_{j-1}}{2 \cdot \Delta x} \right) - \frac{1}{3} \cdot \left( \frac{u_{j+2} - u_{j-2}}{4 \cdot \Delta x} \right) + O(\Delta x)^4$

is fourth order accurate because now the $\partial^3 u/\partial x^3$ terms cancel.

One could continue this procedure but the computer programming would get more
complicated and the treatment near boundaries becomes more complicated too.
For example, if we use the centred fourth order scheme for $\partial u/\partial x$ the first
two points near a north-south boundary must be given special treatment.

Likewise, higher derivatives such as $\partial^2 u/\partial x^2$ can be approximated. A second
order accurate approximation can be obtained by taking

$$(u_{j+1} - u_j) - (u_j - u_{j-1}) = u_{j+1} + u_{j-1} - 2u_j = \frac{\partial^2 u}{\partial x^2} \cdot (\Delta x)^2 + O(\Delta x)^4 \ .$$

Higher order approximations can be obtained using more points.

Given that we can calculate first and second order *spatial* derivatives we can
then approximate the *temporal* derivatives in the equations to follow the time
evolution in our model. The simplest approach is to use a forward difference approximation, using values at the past time step to calculate the
spatial gradients, that is $(u_{n+1} - u_n)/\Delta t$ = a function of spatial derivatives
at time $t = n \cdot \Delta t$, where $\Delta t$ is the time step. (Here the subscripts indicate
the time at which we evaluate u at each point. For simplicity we have omitted
a second subscript indicating the spatial location.) Such a scheme is first
order accurate in $\Delta t$. A scheme which is second order accurate (time errors
proportional to $(\Delta t)^2$) is the leap frog scheme $(u_{n+1} - u_{n-1})/2 \cdot \Delta t$ = function
of derivatives at time $t = n \cdot \Delta t$. This scheme is more accurate but involves
storing values in the computer at three time levels rather than two.

These schemes where the values at new times are calculated from values at
previous times, are called explicit. One can proceed from grid point to grid
point calculating values for the next time step. It is also possible to
devise schemes in which the values at the new time step depend on spatial
gradients of values at the new time step. One then writes down equations
for each grid point and must solve the whole set of linear equations simultaneously. Such methods are called implicit. Because for a large grid, a large
system of equations must be solved one may require more high-speed storage
area in one's computer for such a scheme than for explicit methods.

There are many ways in which the equations of motion and continuity can be
formulated in finite difference form. The question arises as to whether
these various formulations may be distinguished on grounds of relative merit.
Broadly speaking, the assessment of a particular scheme is carried out by
comparing the solutions obtained using a finite difference representation of
the linearised equations of motion and continuity with known simple analytical

solutions.  In this manner there emerge a number of criteria crucial to the
ability of the numerical model to simulate oceanographic phenomena correctly.
Thus, for example, there exists the problem of linear computational instability.  In the case of straightforward explicit schemes there is an upper limit
on the time step which is determined by the distance step and the maximum
wave propagation speed $C_{max}$ in the model according to the relation
$\Delta t \leq \Delta x/C_{max}$.  If this time step is exceeded there results an explosive
growth of small errors inevitably present in the numerical operations.  Implicit methods seem to have better computational stability and it may be
possible to use larger times steps than those given by the condition above.
However, if one makes the time step too large it may be that the implicit
solution obtained, while it is stable, is not a true solution.

Further problems can occur in that the amplitude and speed of a wave in the
numerical model can differ significantly from that in nature, and further
that spurious oscillations may be introduced which are entirely of computational origin.  Instances of the latter have been encountered, for example,
in schemes using central differences in time.  In addition, non-linear terms
such as the advective accelerations may lead to non-linear instability, e.g.,
a rapid spurious increase of energy at small scales which is unrelated to any
real physical phenomenon.  In practice, various techniques have been developed
over the years to deal with these numerical problems.  It is important to
emphasize, however, that a particular model is limited to a certain range of
motions within the sea which it will simulate satisfactorily.

Further discussion of these aspects of numerical modelling is beyond the scope
of this book.  Here we have tried to present an outline of the approach used;
the reader desiring further information should consult the literature.  The
symposium proceedings on Numerical Modelling (Reid, 1975) given in the Further
Reading list would provide a good starting point.

## GENERAL APPROACH TO NUMERICAL MODELLING OF OCEAN CIRCULATIONS

The equations of motion, continuity and, where applicable, heat and salt conservation, are put in finite difference form.  All terms including the non-
linear ones, can be incorporated simultaneously, a real advantage over analytical methods.  Some suitable grid in two or three dimensions is chosen.
Boundary conditions, for example wind stress, temperature and salinity at the
sea surface, and temperature, salinity and velocity on lateral boundaries, are
chosen.  Initial conditions at time $t = 0$ must also be chosen.  Often a state
of rest, that is all velocities initially zero, is assumed.  The temperature
and salinity values may be taken to be uniform in the interior but it is more
usual to prescribe a depth distribution approximating that in the real ocean.
The calculation then proceeds by stepping forward in time, one step at a time
and the process is spoken of as time 'integration'.

As the spatial resolution is increased (i.e., separation between grid points
is reduced) the time step must be decreased to maintain computational stability, at least for the explicit techniques that are usually used.  Thus, for
a two-dimensional grid, doubling the resolution probably requires an eight-
fold increase in the computational time; in three dimensions if the number of
vertical levels is also doubled the amount of work goes up sixteenfold.  Thus
it is easy to see why a desire to improve the resolution of a model may be
limited by the speed and storage capacity of the computer.

Sufficient friction is generally required for computational stability in these non-linear models, either built into the numerical scheme or included in the equations being used. Nature works this way too, of course; as the motion becomes stronger, frictional effects increase until there is a balance between the rate at which energy is being put into the motion and the rate at which it is being lost through friction. Because energy sources are finite the motion remains bounded although it may be quite strong and have large variations in both time and space.

As discussed in Chapter 7 for momentum and Chapter 10 for heat and salt, the system of equations is not closed. Some additional equations must be incorporated to relate the frictional effects to the calculated large-scale velocity field, and the diffusive effects to the gradients of the temperature and salinity fields. These effects are produced in some turbulence-like fashion by motions on scales too small to be resolved by the grid. These are called sub-grid scale phenomena. To close the system the effects of these scales are usually parameterized in a simple way through the use of eddy viscosities and diffusivities. Typically, constant values are used (with different values in the horizontal and vertical, of course).

The eddy viscosity and the grid spacing or the resolution of the model are related to some extent. As noted, friction must be kept at a reasonable level and at the same time velocity differences between adjacent grid points must not become too large. Thus if we use larger grid spacing the maximum velocity gradient is limited. Because the frictional stress is taken as equal to the eddy viscosity times the velocity gradient (e.g., equation 7.4), then to maintain friction approximately constant the eddy viscosity must be increased as the grid spacing is increased. Because the resolution is limited by the speed of available computers the eddy viscosities used are often larger than inferred on the basis of our limited observations and may prevent the non-linear terms from playing a proper role.

Recently in some atmospheric models, variable eddy viscosity has been adopted. It is made proportional to the root mean square rate of strain in a calculated flow (the rate of strain is derived from the calculated shears). Physically, this variable viscosity with friction being larger where the shears or velocity gradients are larger seems a better representation than constant eddy viscosity and it seems to improve the atmospheric models. As yet this approach has had little use in oceanic models. (O'Brien's model, described later, is one exception.) Also, the variable eddy viscosity seems to damp scales of motion near the grid scale more strongly relative to larger scales than constant eddy viscosity. Thus, computation stability can be maintained with lower, more realistic average viscosity values. The use of variable viscosity may be helpful in allowing non-linear effects to be more realistic with the resolution capabilities of present computers.

As noted above, one of the boundary conditions to be imposed is the forcing by the wind stress. As discussed in Chapter 9, the procedure for calculating the wind stress from the observed wind velocities is still not well established, particularly for stronger winds. In the models to be described in the next section the imposed wind stress is usually based on computations made by Hellerman. These computations are based on climatological wind data and have been carefully done in an attempt to take into account the variations in wind speed and direction at a given location. The drag coefficient formulation that Hellerman used in these computations was the smooth version of the step function with values for $C_D$ of about $0.8 \times 10^{-3}$ at low wind speed rising

rapidly but smoothly in the vicinity of 7 m/sec to 2.4 x $10^{-3}$. It was known
at the time (about 1965) that Hellerman did the calculation that the drag
coefficient does not have this rapid transition, but Hellerman was not aware
of the more recent results. Thus this calculation may have overestimated
somewhat the contribution of strong winds and also perhaps underestimated
the contributions in regions of light winds, that is in the regions in the
middle of gyres where the stress changes sign and the curl of the wind stress
is a maximum, although even here, climatologically, winds less than 6-8 m s$^{-1}$
are unimportant. The spatial resolution of Hellerman's stress values is also
limited because of the way the climatological data are compiled as described
in Chapter 9. Because of this limited resolution the stress gradients or the
curl of the wind stress may also be underestimated. It is difficult to know
how serious these errors are but it is not likely that the true values of the
maximum of the curl are more than about 50% larger than Hellerman's calculations. Some more detailed calculations for the regions of maximum curl in
the North Atlantic, which have better spatial resolution and use more up-to-
date drag coefficient values, give similar values for the maximum curl to
those obtained using calculations similar to those of Hellerman. However,
these calculations produce a long-time average and do not take into account
seasonal variations in the position of maximum curl. As noted in Chapter 9
such a calculation may give a lower value than if we calculated a maximum curl
on a daily, weekly or monthly basis and then averaged these maximum values
together regardless of the latitude of the maximum curl. It is the latter
value of the maximum curl which is probably most appropriate for calculating
the maximum transport due to the wind using the simplified Sverdrup equation
(9.21).


## DESCRIPTIONS OF SOME MODELS OF INDIVIDUAL OCEANS

In this section we shall outline the features and main results of a few numerical simulation models to show the sort of results which can be achieved and
also some of the limitations. We start with a fairly simple model and proceed
to somewhat more detailed models culminating in a description of two models
of the whole world ocean.

### O'Brien's Two-Dimensional Wind-Driven Model of the North Pacific

This is a rather simple vertically averaged model similar to the classic Munk
analytical model described in Chapter 9, except that it uses more realistic
geometry and includes non-linear effects. It has the following features:

1.  Density is taken to be uniform, that is, the model is barotropic; the vel-
    ocities are independent of depth, reducing the problem to two dimensions.

2.  It is driven by the wind using the climatological average wind stress
    calculated by Hellerman.

3.  The equations are in spherical polar form, instead of using the beta plane
    approximation as Munk did.

4.  The 'grid' is 2° in latitude and longitude.

5.  Bottom topography is included above 2000 m, the assumed depth of the
    ocean.

6.  Lateral friction is used with a variable eddy viscosity as in some atmos-
    pheric models - the only ocean model that we know of which includes this
    feature.

Fig. 11.1.   Mass transport stream function ($\psi$) computed for a wind-driven
            model of uniform density (North Pacific): (a) the circulation
            pattern at the western boundary in units of $10^6$ tonnes $s^{-1}$
            $\simeq$ Sverdrup ($10^6 m^3 s^{-1}$), (b) the pattern over the entire basin.
            (From J.J. O'Brien, *Invest. Pesqu.*, 35, 341, 344, 1971.) (See
            note in the text on the signs for $\psi$.)


The model is started from a state of rest. As the integration proceeds, flow
begins to occur and to build up into a series of gyres. This process is
spoken of as 'spinning up'. After some time a fairly steady circulation is
apparent but there are fluctuations superimposed upon it. The transport
pattern after 72 days of integration is shown in Figure 11.1. Figure 11.1a
shows the detailed structure at the western boundary and Figure 11.1b the
broad-scale pattern over the whole ocean basin. Note that the sign convention
for the stream function in this figure is opposite to that which we adopted in
Chapter 9. This difference is present in all the figures in this chapter, so
that the modellers whose work is reported here are all consistent but contrary
to the usual fluid mechanics convention. In this chapter, a gyre with a
positive $\psi$ value at its centre indicates a clockwise gyre. There are arrows
in some of the figures but in any case the reader should know the directions
of the main circulations by now and be prepared for the lack of consistency
in the literature.

The model seems to produce known features of the circulation reasonably well,
although there are differences in detail. The Kuroshio leaves the coast at
35°N and has a transport of about 60 Sv in reasonable agreement with observations. Earlier estimates from geostrophic calculations suggested a transport
of 65 Sv but more recent estimates are somewhat higher, perhaps 80 to 90 Sv.
Munk's calculations gave a transport of 38 Sv and the current did not leave

the coast until about 45°N.  The numerical model thus seems to give better
results.  The departure from the coast at 35°N is partially an effect of the
topography.  (Topographic effects may be overemphasized in barotropic models
as we shall see.  However, in this case, the bottom is taken to be flat for
real depths greater than 2000 m so the problem is probably not serious in
this model.)

The greater transport, compared to Munk's linear model, is probably due to
enhancement of the flow by non-linear effects (the advective acceleration
terms).  The early numerical models mentioned previously which were used to
investigate the effect of the non-linear terms showed quite clearly that the
flow could be enhanced by these effects by comparison with the results obtained when the non-linear terms were omitted.  Thus, as Stommel suggested, non-
linear or inertial enhancement can lead to stronger circulation.  Much of the
additional transport is a local recirculation, that is, the additional flow
in the western boundary current is mainly returned in a fairly strong counter-
current close to the boundary current.  O'Brien's result (Fig. II.I) shows
this behaviour with about 1/3 of the Kuroshio transport associated with local
recirculation.  The southward flow in the interior for the main gyre is about
40 Sv, just what one would calculate using the Sverdrup equation as must be
the case for a model with a flat bottom, provided that the non-linear and
bottom friction effects are unimportant in the interior as we expect to be the
case.

As we shall see, other numerical models of the North Pacific of similar
resolution, with constant eddy viscosity and with the same wind stress, give
transports more like that of Munk's study.  It appears that the variable eddy
viscosity allows lower overall viscosity and hence more important non-linear
effects, and leads to enhancement of the transport of the Kuroshio to more
realistic values.

While O'Brien's model seems to produce results fairly consistent with our
limited knowledge of the actual circulation, the maximum transport values
produced are still somewhat lower than the most recent estimates although
there is uncertainty both in these transport estimates and in the wind stress
O'Brien used in his calculations.  However, the assumption of constant density
leaves out any thermohaline circulation completely.  This separation has
always been necessary in analytical work because the whole system cannot be
treated at once.  In addition to the non-linear nature of the equations
describing the velocity field, there is a fundamental non-linearity - the
advective terms in the equations for salt and heat conservation as discussed
in Chapter 10.  It seems most unlikely that the flows produced by the two
driving forces do not interact, which provides, as we shall see, another
possible enhancement mechanism, in addition to the simple linear addition of
the thermohaline flow to the wind-driven flow that was originally suggested by
Stommel.  O'Brien's model shows the effect of inertial enhancement in a
simulation model and how Munk's approach works when treated more realistically.

Cox's Model of the Indian Ocean

This model is an attempt to see if the current reversals in the Indian Ocean
that occur as the Monsoon winds change can be reproduced in a numerical model.
The model is three-dimensional with up to 7 levels in the vertical depending
on the depth.  It incorporates the following features:

1.  Wind driving is seasonal, based on three-month averages over each of the
    four seasons using the Hellerman wind stress.  The stress is made to vary
    smoothly from season to season.

2.  Thermohaline driving is included by imposing the surface temperatures and
    salinities.  While surface fluxes of heat and salt would be preferable,
    the surface salt and temperature values are imposed because the fluxes
    are not nearly as well known.  Temperatures are imposed with smooth varia-
    tion from winter to summer values.  The salinity field is held constant
    because the seasonal variation of the surface salinity appears to be small.

3.  The 'rigid lid' approximation is imposed as is quite common in ocean
    models (O'Brien's model is an exception).  This approximation is achieved
    by requiring that the vertical velocity at the surface be zero.  The
    effects of real variations of sea surface elevations appear as the equi-
    valent pressure distribution on the rigid, level, upper surface of the
    model.  The purpose of this approximation is to allow longer time steps
    to be used; with a rigid upper surface, surface gravity waves which have
    rapid propagation speeds are not allowed.  Rossby waves in which the
    variation of the Coriolis force with latitude provides the restoring
    force (see Chap. 12) are still permitted but are modified somewhat by the
    presence of a rigid lid instead of a free surface.  It is believed that
    this modification produces no important errors.  It may modify the initial
    'spin up' processes but probably produces no serious errors in the final
    quasi-steady state.

4.  The model is started from a state of rest, the initial salinity and
    temperature values are taken to be horizontally uniform but with a vertic-
    al distribution based on observed data.  On open boundaries salinity and
    temperature are specified at all depths using observed data.  The baro-
    tropic (that is depth independent) and baroclinic (depth varying) flows
    are calculated separately subject to no net vertically integrated flow
    through the open boundary to the south.

5.  The friction and diffusion of heat and salt are parameterized with con-
    stant eddy viscosity and diffusivity.  $A_z$, eddy viscosity for vertical
    friction, = $10^{-2}\,m^2\,s^{-1}$ for the top layer (50 m) and = $10^{-4}\,m^2\,s^{-1}$ for the
    remaining layers.  $A_H$, eddy viscosity for horizontal friction, is large
    at first, $2 \times 10^5\,m^2\,s^{-1}$, but is finally reduced to $5 \times 10^3\,m^2\,s^{-1}$ as the
    grid is refined (as described below).  Vertical eddy diffusivity is taken
    to be constant at $10^{-4}\,m^2\,s^{-1}$ throughout; horizontal eddy diffusivity is
    initially $10^4\,m^2\,s^{-1}$ and finally $5 \times 10^3\,m^2\,s^{-1}$.

6.  The density field is not allowed to be statically unstable.  Whenever
    such a state is predicted in a new time step it is assumed that vertical
    mixing occurs immediately to produce a neutral density structure with
    heat and salt conserved.  This mixing hypothesis is commonly used to avoid
    the problem of static instability in numerical models of the ocean (and
    of the atmosphere).

7.  Because of the long response time of the density field in the interior
    (of the order of 200 years) the calculations are done in three stages
    First, bottom topography is ignored and computations are carried out for
    a 4° grid.  The integration proceeds for 130 years of model time.  (It
    takes 0.2 hours of computer time per year of model ocean time.)  In the
    next stage the grid is reduced to 2° and the integration proceeds using
    the values from the first stage as a starting point.  Bottom topography
    is included in this stage and the integration proceeds from 130 to 185
    years in the model (at 1.7 hours of computer time per year of model time).

Fig. II.2.    Cross-sections showing the long-shore velocity component (cm s$^{-1}$)
near the Somali coast in an Indian Ocean model: (a) during the
south-west monsoon (May-September), (b) during the north-east
monsoon (November-March). (From M.D. Cox, *Deep-Sea Res.*, 17,
68, 71, 1970).

In the third and final stage which proceeds using the second stage as a
starting point the grid is reduced to 1°, the surface layer is split into
two layers and integration from 185 to 192 years is done (at 22 h/year).
Thus the total computation requires about 270 hours of computer time on a
Univac 1108.  Refinement to a 0.5° grid would require 200 to 300 hours of
computer time per year of model time.  Faster computers have now become
available but at the time Cox did his work, resolution better than the 1°
of his final stage was clearly impractical.

The general features of the solution for the final year of the computation
agree fairly well with the rather sparse observations.  A Somali current
(western boundary) appears in the model and shows the proper seasonal variations.  Figure II.2 shows the longshore component of the velocity in vertical
sections normal to the Somali coast.  The Somali current is well developed in
the northern summer solution and absent in the northern winter.  One cannot
compare the calculated transport in the Somali current with an observed value
because of insufficient field data.  This example illustrates the serious and
typical problem of insufficient verification data.

Although the transports and general features of the density field look reasonable the computed velocities in the Somali current are too small and the horizontal density gradients are not as large as in nature.  These discrepancies are probably due to using eddy viscosity and diffusivity values that are larger than they should be, causing the boundary current to be 'smeared out'. To see if these effects are serious in the sense that the solution is not correct to the first order (that is it is not a smoothed version of what one would obtain with finer resolution but something quite different because, for example, of suppression of non-linear terms) the calculation would have to be continued at finer resolution.  With the new, faster computers that have now come into use, such calculations will be feasible.

While the seasonal response seems to have been correctly reproduced here, the question of whether or not models with large eddy viscosities and diffusivities produce results correct to first order is far from simple.  For example, in a model of the North Atlantic done by Holland which uses the same value for the horizontal eddy diffusivity of heat as the Cox model, the vertical circulation is quite different from what we expect.  The model produces *downwelling* in the vicinity of the main thermocline, but it is generally believed (although with no good observational proof) that there is upward flow over most of the ocean at this level.  This assumption was made by Stommel in his abyssal circulation models which seem to be consistent with the limited observational data, although as discussed in Chapter 10 the way in which the thermocline is maintained is not clear.  Such a difference in the vertical flow field requires a modification of the horizontal flow to maintain the thermocline.  In a later calculation with a smaller diffusivity ($10^3 \, m^2 \, s^{-1}$) upwelling in the interior was obtained in agreement with Stommel's hypotheses. Because of the large horizontal temperature gradient in the western boundary region in Holland's model, the large horizontal diffusivity produces strong diffusion normal to the potential density surfaces.  The effective diffusivity across the surfaces is about $5 \times 10^{-3} \, m^2 \, s^{-1}$ in an essentially vertical direction.  To balance this strong effective vertical diffusion, strong upward advection is required in the western boundary.  By continuity, downwelling is forced in the interior.  Reducing the diffusivity reduces the effect and upwelling in the interior is then found in the model as is believed to be the case in nature.

This example illustrates that the values of eddy viscosity and diffusivity which must often be chosen for computational stability rather than to agree with values inferred from observational data may be critical to obtaining realistic results.  The large values of eddy viscosity which are used probably do not produce such dramatic effects because the surfaces of constant velocity are less likely to have tilts comparable to those of the isothermal surfaces. If the surfaces of constant velocity were tilted then the same sort of effect would occur and the effective vertical diffusion of momentum might be stronger than that in reality, leading to greater coupling of the layers in the vertical than is realistic.  To answer such questions requires running the models with finer resolution and lower viscosity and diffusivity to see if the results are different from those from the coarser resolution models.

Holland and Hirschmann's Model of the Atlantic Ocean

The computational scheme used in this model is similar to that of the Cox model of the Indian Ocean.  However, unlike the Cox model it is what is called a *diagnostic* model rather than a *predictive* or *prognostic* model.  The density field is prescribed, that is it is based on observed data and is held fixed rather than predicted as part of the calculation. The North Atlantic is chosen because it is the area of the world ocean with the best observed density data, and also the best knowledge of the circulation, although our knowledge is still limited, even here. The great advantage of the diagnostic model is that it takes a small fraction of the computer time of a prognostic model. The model ocean reaches its state of statistical equilibrium in about a month instead of several hundred years. Thus, more parameters can be varied and finer resolution can be used than in a prognostic model.

The features incorporated in this model are:

1.  Horizontal resolution of 1° in latitude and longitude and 14 levels in the vertical.

2.  The basin extends from 11.5°S to 50.5°N, including the equatorial region, to test the model there.

3.  Climatological density data are used, smoothed to the 1° grid.

4.  Climatological wind-driving, based on the Hellerman wind stress.

5.  Vertical eddy viscosity is taken to be $10^{-4}$ $m^2$ $s^{-1}$; some tests have indicated that this value is not critical in a diagnostic calculation.

6.  The horizontal eddy viscosity is $4 \times 10^4$ $m^2$ $s^{-1}$. (Some calculations were attempted with the smaller value of $10^4$ $m^2$ $s^{-1}$ but eddies with size comparable to the grid spacing began to appear. To prevent accumulation of energy in eddies of the smallest size that can be resolved, which could lead to non-linear computational instability, the larger value of eddy viscosity is required.)

This model seems to produce the main features of the circulation rather well. However, the surface currents of the equatorial region and the region south of the equator appear rather more broken up than observations suggest. This characteristic of the model is perhaps due to insufficient resolution and to inadequate density data. Transport through the Florida Straits is much smaller than observations indicate because of the limited resolution. The Gulf Stream gyre, both the surface circulation and the transport (maximum value 81 Sv), looks reasonably correct although the Gulf Stream is broadened and the velocity values are low due to the large eddy viscosity. Recent observations based on direct measurements suggest maximum transports of 100 to 150 Sv so the calculated value is still below what the observations indicate although it is over twice that of the classical linear theory of Munk.

The calculated Gulf Stream transport is much larger than the Sverdrup transport due to the wind (calculated from the simplified Sverdrup equation 9.21). This transport would be about 25 Sv, similar to but somewhat smaller than the value obtained by Munk. It is of course based on the Hellerman wind stress and may be somewhat lower than it should be, but even doubling the value, which is probably the outer limit, would not produce a balance between the Sverdrup transport and the Gulf Stream transport.

The reason for the enhancement of the Gulf Stream in this model has been quite clearly demonstrated by doing computations with simplified versions of the model. Figure 11.3 shows the transport stream function for three cases. Figure 11.3a shows the case in which the density is taken to be constant, all other features of the model being unchanged. Thus, this is the barotropic case but with the real bottom topography within the resolution limits of the model. The Gulf Stream transport for this case is only 14 Sv. The bottom

topography has reduced the flow below what one would expect using the simplified Sverdrup equation (9.21).  As discussed in Chapter 9, when the current

Fig. 11.3.  Transport stream functions (in Sverdrups) for the Atlantic Ocean
from diagnostic calculations based on the observed density field:
(a) for uniform density, (b) for the observed density field but
uniform depth of 1,273 m, (c) realistic case of observed density
field and bottom topography, allowing bottom pressure torques.
(From W.R. Holland & A.D. Hirschmann, *J. Phys. Oc.*, 2, 342, 343,
348, 1972.)

extends to the bottom and the bottom is not level, an additional term has to
be added to the simplified Sverdrup equation.  As is to be expected, topographic steering of the currents is very important and is probably much
stronger than in reality.  In Figure II.3b the transport pattern is shown for
the case in which the observed density field is used but the bottom is assumed
to be level at a depth of 1,273 metres.  The transport pattern is quite smooth
and much like the results of the classical Munk theory except for a little
distortion because of the more realistic geometry.  This result could have
been anticipated since the model is now essentially the same as the Munk
model except for the more realistic coasts.  The transport is virtually the
same as that calculated using the simplified Sverdrup equation for the interior transport as expected.  Figure II.3c shows the transport stream function
for the case in which all the features of the model are included.  The transport pattern has become quite complicated and in addition the maximum transport in the Gulf Stream has increased to 81 Sv as noted earlier.  This enhancement of the Gulf Stream is produced by what has been called the joint effect
of baroclinicity and bottom topography.  Baroclinicity and thermohaline
driving are required to produce  deep currents that interact with the bottom
topography through the term involving the deep current and the bottom slopes
in the case of a non-level bottom.  A purely barotropic case is not correct
because topographic effects reach to the surface instead of modifying the deep
flow, which then interacts with the upper layer.  The interaction of the deep
currents with the sloping bottom produces pressure torques which can accelerate the flow and put vorticity into the flow just as the wind stress curl
does.  In the model this effect can be larger than the wind stress curl in
some regions, particularly in the Western Atlantic.  This effect has also
been demonstrated in a mechanistic model with idealized geometry by Holland,
and in a diagnostic model of the North Atlantic done by Sarkisyan and Ivanov.
There is an analog of this effect in the atmospheric flow, where the presence
of mountains can, through pressure torques, add vorticity to the flow.

Unfortunately, how important this effect is in the real ocean is not certain.
The possible errors in the input density data prevent the diagnostic calculations from being conclusive.  These errors arise both from instrumentation
limits and from the difficulty of obtaining good values of a time-varying
field.  It seems that there is enough evidence for the possible importance
of this mechanism to justify a carefully planned observational experiment.

The transport pattern produced by the full model is rather complicated, much
more so than we expect from the classical pictures of Chapter 9.  However,
we don't know, in fact, whether the real transport pattern is like the one of
the model or not.  Our knowledge of the large-scale flow of the ocean below
the surface is based almost entirely on indirect evidence either from trying
to guess what the flow must be like to produce the observed property distributions, or from geostrophic calculations which are always uncertain because
of lack of knowledge of the level of no motion.  The only direct observations
of currents that we have for large regions are of the surface drift.  The
surface drift values are based on ships' navigation records and any individual observations may have quite a lot of error, so a great deal of averaging
and consequent smoothing is done in the process of producing charts of these
currents.  In spite of the complicated transport pattern, the results of the
Holland and Hirschmann model are not inconsistent with our limited observational knowledge.  Figure II.4a shows the surface pressure distribution (as
$(p/\rho_0)$ for the full model.  The model is a rigid lid one, so this surface
pressure distribution if divided by g would be the equivalent surface elevation.  Although it is slightly smoother it shows strong similarities to the

Fig. 11.4.    For the diagnostic model of the Atlantic Ocean: (a) the surface
              pressure $(p/\rho_0)$ calculated from the results for Fig. 11.3c
              after reaching a steady state (after 35 days), contour interval
              $1\,m^2\,s^{-2}$ (0.1 dyn m), (b) the dynamic topography based on a 1,000 m
              reference level, contour interval $1\,m^2\,s^{-2}$.(continued)

Fig. 11.4.    (continued)   Diagnostic model of the Atlantic Ocean: (c) horizontal velocity vectors for the surface layer; the lengths of the
vectors in the Gulf Stream and equatorial regions are limited to
a maximum of $0.75\,\mathrm{m\,s^{-1}}$ in order not to obscure smaller scale
features, the largest velocity being $1.1\,\mathrm{m\,s^{-1}}$ in the Gulf Stream
(from W.R. Holland & A.D. Hirschmann, *J. Phys. Oc.*, 2, 345, 346,
1972).

dynamic height based on a 1,000 m reference level, which is shown in Figure
11.4b.   Although in the model 1,000 m is not a level of no motion the dynamic
height variations at this level are small compared with those of the surface,
so the model pattern of equivalent surface elevation is very similar to that
of the dynamic height computation.   It is worth noting that the dynamic topography is not as smooth and simple as the classical pictures that we showed
in Chapter 9.   Except near the equator the flow is nearly geostrophic and so
is more or less along the contours and the Gulf Stream is quite evident in
both Figure 11.4a and b.   Finally, Figure 11.4c shows the surface current
calculated from the model and except for the equatorial region it is fairly
similar to the currents shown on pilot charts based on ships' records.

The large eddy viscosity (near the upper end of the likely range) is a problem
in a model such as this.   Non-linear effects are undoubtedly suppressed.   In
addition, the resolution is insufficient to allow for the presence of mesoscale eddies, recently discovered to exist in the ocean, which may be important to the dynamics of the large-scale flow.   It would be interesting to
know if either inertial enhancement or the effects of mesoscale eddies or
some combinations are large enough to give the additional enhancement of the
Gulf Stream needed to match the observations.

## TWO MODELS OF THE CIRCULATION OF THE WORLD OCEAN

The first by Bryan and Cox is a constant density or barotropic model. The
second by Cox is an extension of the first model to the diagnostic baroclinic
case (temperature and salinity based on observed mean values and held fixed)
and finally to a short period (2.3 years of model ocean time) of a fully
predictive baroclinic case using the final state of the diagnostic model for
initial conditions. It was not possible to run the predictive case to final
equilibrium because this calculation would have required several centuries of
ocean time at 10 hours of computer time per year of ocean time. However, the
initial adjustment in boundary currents to removing the imposed density field,
which takes about a year, can be examined.

The models have the following features:

1.  Realistic topography and coasts within the resolution limits of the
    models.

2.  In the baroclinic case, up to 9 levels depending on the depth.

3.  Horizontal eddy viscosity of $8 \times 10^4 \, m^2 \, s^{-1}$; with the resolution that can
    be used in this model, smaller values of eddy viscosity lead to small-
    scale computational noise.

4.  Vertical eddy viscosity of $10^{-4} \, m^2 \, s^{-1}$.

5.  Wind stress based on the Hellerman computation with extension to the
    Polar regions.

6.  Resolution of 2° from 62°S to 62°N. Separate 2° spherical grids are used
    for both higher latitude regions with considerable overlap between these
    grids and the lower latitude grid. Each grid is integrated separately
    for several time steps and the boundary conditions are updated in the
    overlap region. This approach is required because, on a full spherical
    grid, the grid spacing becomes small near the poles requiring unreason-
    ably small time steps everywhere.

7.  No conditions on open boundaries (fluid boundaries within the ocean's in-
    terior) are required as all lateral boundary openings are included in a
    world ocean model. Such boundary conditions are always somewhat arbi-
    trary due to a lack of adequate observations and are a problem when
    attempting to model a part of the world ocean.

The transport pattern for the barotropic case is shown in Figure 11.5a. Topographic steering effects on the current appear to be much stronger than they
should be; western boundary currents such as the Kuroshio and Gulf Stream are
too weak. The Antarctic Circumpolar Current is particularly weak (about
21 Sv). In a separate calculation (with the different horizontal eddy viscosity of $4 \times 10^4 \, m^2 \, s^{-1}$) the barotropic model was run with a level bottom and
the Kuroshio and Gulf Stream then had transports similar to those calculated
by Munk, as can be seen in Figure 11.5b. As can also be seen in this figure
the Antarctic Circumpolar Current is very large (greater than 600 Sv and still
increasing slowly at the end of the calculation).

In the North Pacific, the level bottom case is almost equivalent to O'Brien's
North Pacific model shown in Figure 11.1, except for the difference in eddy
viscosity. The difference in the strengths of the Kuroshio Current in these
two models provides evidence for our earlier suggestion that inertial enhancement produces the larger transport in the O'Brien model.

Fig. 11.5.    World ocean circulation: (a) horizontal mass transport stream
function for uniform density with realistic topography (positive
values indicate clockwise flow), Antarctic Circumpolar Current
$= 22 \times 10^6$ tonnes $s^{-1} \simeq 21$ Sv. (from M.D. Cox, *Num. Mod. Oc.
Circ.*, p. 111, 1975.) (b) Transport streamlines for uniform
density and depth with much larger Antarctic Circumpolar Current
($\sim 600$ Sv) (from K. Bryan & M.D. Cox, *J. Phys Oc.*, 2, 326, 1972.)

Fig. 11.6  World ocean circulation: (a) pattern of mass transport for the
           diagnostic case based on the observed density field, (b) pattern
           of mass transport for the predictive case based on a 2.3 year
           numerical integration of a three-dimensional model using observed
           density and the mass transport of Fig. 11.6a as initial conditions.
           (From M.D. Cox, *Num. Mod. Oc. Cir.*, pp. 112, 113, 1975).


The transport pattern from the baroclinic diagnostic model by Cox is shown
in Figure 11.6a.   The results appear to be much more realistic with more
reasonable transport values for the western boundary currents, although they
are still smaller than observed values and the currents are broader and weaker
because of the limited resolution and large eddy viscosity.   The Antarctic
Circumpolar Current appears more reasonable too, with a transport value of

$184 \times 10^6$ tonnes $s^{-1} \simeq 180$ Sv, near the upper end of the range of values suggested by observation. Again, the inclusion of baroclinicity and bottom topography has enhanced the western boundary currents to levels beyond what is expected from the simplified Sverdrup equation (9.21). The enhancement (to about 60 Sv) is not as large in the Gulf Stream as found in Holland and Hirschman's model, perhaps because of the larger viscosity and lower resolution. Many comparisons may be made between the model and observed features of the ocean; good agreement is obtained in some regions and not in others. The model is only a first step toward a model representative of the world ocean, but it is an important one.

The transport pattern for the predictive model is shown in Figure 11.6b. There is a smoothing out of various details, probably due in part to inaccurate density values; in addition the large eddy viscosity and diffusivity which must be used probably lead to oversmoothing and to the reduction in the transport values of the Gulf Stream and Kuroshio.

## MODELS OF MESOSCALE EDDIES

Recently a considerable amount of evidence has been collected for the existence of *mesoscale eddies* in the ocean. These eddies seem to have characteristic sizes of the order of 200 to 500 km, time scales of 1 to a few months and kinetic energies between 10 and 100 times that of the mean flow in the interior. Most of the energy associated with the mean flow is the potential energy of the tilted isobaric surfaces; this energy is perhaps as much as 1,000 times the kinetic energy. However, the potential energy of the eddies is more comparable to the kinetic energy, so the total energy of the mean flow is probably larger than the total energy of the eddies even if they are found everywhere. How prevalent these eddies are and their importance to the mean flow is not established as yet. The atmospheric analogues of the oceanic eddies are the storm systems in mid-latitudes. On a weather map they look like a series of large eddies. These eddies can be considered to be geostrophic turbulence which behaves like two-dimensional turbulence. In the atmosphere it is known that the eddies gain energy from the available mean potential energy and transfer kinetic energy and momentum to larger scales and to the mean flow. They are important in determining the strength of the westerlies and the jet stream, for example. If one tries to parameterize the effects of these eddies by an eddy viscosity in the usual simple way, the value is negative because they accelerate the flow in contrast to the more familiar smaller scale three-dimensional turbulence which acts to retard the flow. If the oceanic eddies are fairly common and have similar dynamic behaviour to their atmospheric counterparts they may be important in the dynamics of the mean flow.

The first evidence for these eddies that was taken seriously was obtained by Swallow when he first used his floats in an attempt in 1959-60 to observe the expected very slow flow in deep water. Indeed, more recent but very limited observations suggest that if he had gone further into the interior he might have found the quiet ocean which he expected - there do seem to be some quiet regions. In any case, much to his surprise he found that his floats went off in various directions at speeds at least 10 times greater than expected, so that he was unable to follow them for long. Once one accepts the idea that eddies are a feature of the ocean, one can find lots of evidence in the historical records. While much of the early data had station spacing too large to show the eddies clearly, there were some detailed observations. In these

the eddies can be seen in the density field (and in the temperature field as
the dominant factor determining density in the open sea) because the flow
around the eddies is nearly geostrophic. (As in western boundary currents
and atmospheric weather systems, the Coriolis force associated with the along-
stream component dominates the cross-stream momentum equation.) In many ways
it is curious that oceanographers did not consider the possible importance of
eddies and look for them sooner. Everyone is well aware of the meteorological
analogue - the weather may easily make a pleasant climate unbearable. Oceanographers continued to try to find and understand the ocean climate (the mean
state) perhaps rather longer than they should have. Eddy effects were mis-
interpreted as internal wave noise or observation errors or were just
ignored.

Up to now it has not been possible to use fine enough resolution in a simulation model to allow for the presence of mesoscale eddies. However, mechanistic models with simplified geometry of sufficient resolution have been
examined and have shown that the mesoscale eddies may produce important
effects on the general circulation. For example, they can extract energy
from the potential energy of the mean flow of the upper layer wind-driven
circulation, transfer it to eddies and mean circulation in the deep layer of
the ocean, and in this way increase the total transport of the system.

The instability mechanism which allows these eddies to grow with their energy
supply coming directly from the mean potential energy field is called *baroclinic instability*. Baroclinic flow with vertical shear is required for this
mechanism to be possible. It is different from dynamic instability (Chap. 7)
in which eddies may be generated by shear in the mean flow and gain their
energy from the mean flow kinetic energy. In contrast to baroclinic instability, this shear instability is called *barotropic;* as there is no vertical
shear, horizontal shear is required for disturbances to grow. Both of these
instability mechanisms are important in the atmosphere; their relative importance in the ocean has not been established but as eddies exist either or
perhaps both types of instability are likely sources.

With barotropic instability, energy comes from the mean flow but the mean flow
may extract energy from the potential energy field so that indirectly the
energy for barotropic instability may be supplied from the mean potential
energy field. Rotation, while it does not directly affect dynamic stability
as noted in Chapter 7, plays an indirect role. Because of the importance of
the Coriolis term much larger isobar slopes are required than in the non-
rotating case so that the pressure gradient can balance most of the Coriolis
term. The larger isobaric slopes lead to a much larger mean potential
energy. Thus rotation leads to a large mean potential energy and provides
a much bigger possible energy source for instability than that available in
the non-rotating case.


## COMMENTS ON THE NUMERICAL MODEL SOLUTIONS

Many models use the Hellerman wind-stress calculation although it is not the
best calculation that could be done as discussed earlier in this chapter.
Nevertheless, using the same stress values for different models is useful for
comparison and the inaccuracy is probably not too critical at the present
stage of modelling. However, one cannot make quantitative arguments about
the accuracy of the Sverdrup equation (9.21) using this computation. Hellerman is presently redoing his stress computation with more extensive data and

using a constant drag coefficient.  As noted in Chapter 9 the dependence of the drag coefficient on wind speed and other parameters is not well established as yet.  However, if new measurements should show that a linear dependence on wind speed is a better representation for the drag coefficient it should be possible to scale this new calculation of Hellerman's because at a given location the range of wind speeds which contribute to the stress is fairly narrow.

Some modelling to test the sensitivity of the results to variations in the stress input would be most helpful in determining how well this forcing function needs to be known.  This information would also be useful to those attempting to make better measurements of the drag coefficient.  In diagnostic models this stress function seems to be almost immaterial; the information is in the imposed density field that the wind is partially responsible for setting up.  For example, in the model of Holland and Hirschman, putting the wind stress to zero only lowered the Gulf Stream transport by 5%.

In principle, the advantage of the numerical modelling approach is that all terms in the equation can be included and the topography and coastline can be realistic.  At present, at least in the simulation models, the friction seems too large to allow the non-linear terms to play a realistic role.  Exploration of possible parameter ranges has also been limited by the speed of available computers.  If the mesoscale eddies are as important in the dynamics of the ocean as they are in the atmosphere, then they present a serious resolution problem.  It is known in atmospheric models that 250 km resolution gives rather better results than 500 km resolution.  Then all scales that make important contributions to the total kinetic energy, including the storms or mesoscale eddies, are fairly well resolved.  The scale of these eddies seems to be proportional to the *Rossby radius of deformation*, $\lambda = [g(\Delta\rho/\rho)D]^{\frac{1}{2}}/f$, where g is the acceleration of gravity, $\Delta\rho$ is the density difference between the two main layers of the fluid, D is the thickness of the layer, and f is the Coriolis parameter.  For the atmosphere, $\lambda$ is the order of 1,000 km.  For the ocean, $\lambda$ is of the order of 100 km because both $\Delta\rho/\rho$ and D are smaller.  Thus, it appears that a resolution of order 25 km is required in an ocean model to resolve all energetic scales.  To model the world ocean with such resolution is impossible at present.  However, modelling of limited regions can and no doubt will be done (although the region cannot be too limited because the results may then depend mainly on the poorly known conditions on open boundaries).  All these problems are related to the limitations of the size and speed of available computers; however, the new faster machines that have now become available and new numerical techniques that are being developed should help to overcome them.

As noted at the beginning of this chapter there appears to be a real discrepancy between the observations and the predictions of the linear wind-driven theory presented in Chapter 9.  Simple linear addition of the thermohaline circulation discussed in Chapter 10 probably cannot resolve the discrepancy either.  While the numerical models discussed in this chapter do not provide the final answer, they have suggested a number of mechanisms which might explain the discrepancy.  The three possibilities are: inertial enhancement, bottom topography - baroclinicity, and mesoscale eddies.  Another possibility which has been demonstrated in a mechanistic numerical model is the effect of fluctuations in the wind stress; these fluctuations can produce additional mean currents through non-linear rectification effects.  It may well be that all these mechanisms are important at least in some regions, but how important they are in the real ocean remains to be demonstrated.

Finally a cautionary note about this discrepancy.  Our observations are still
very limited and have considerable uncertainties.  Some recent tests of the
simplified Sverdrup relation (equation 9.21) in the interior, with the wind-
stress curl calculated with more up-to-date drag coefficients and better
spatial resolution, seem to give good results.  The flow was calculated by the
geostrophic method and assumed to be baroclinic.  (In the interior, if the
bottom is level the Sverdrup relation tells us that the net north-south transport is given by curl $\tau_n$.  Any additional thermohaline flow must have zero
net transport over the whole depth - the poleward deep flow is exactly balanced by the equatorward flow in the upper layer.  Of course the bottom is not
level and indeed is quite rough in places (e.g., the mid-Atlantic Ridge) so
the baroclinicity - bottom topography effect may play a role.  In this case,
separation into interior and western boundary regions may be difficult.)  The
calculated geostrophic transport agreed very well with the transport calculated from the wind-stress curl.  When extrapolated for the full width of the
ocean, the sections being taken well within the interior to avoid lateral
boundary effects, the calculated transports also agreed very well with the
transports in the Florida Straits which are 30-35 Sv; it is only much further
downstream that the very large Gulf Stream transports (in excess of 100 Sv)
have been observed.  This test of the Sverdrup relation was done where it
ought to be done, in the interior, as was the fairly successful test in the
eastern equatorial Pacific given in Chapter 9.  Some direct check on the geostrophic calculation is really needed also, of course, but the test described
above is much better than trying to compare the interior Sverdrup transport
to the Gulf Stream transport which may not be sufficiently well observed.
With the possibilities of enhancement due to inertial effects, mesoscale
eddies, baroclinicity-bottom topography interactions and perhaps other
effects, the western boundary region may be rather more extensive than it is
presently considered to be.  The observed Gulf Stream and Kuroshio transports
(directly observed or geostrophically calculated with some direct observations
to fix the level of no motion between hydrographic stations) may be based on
observations which do not extend far enough offshore to detect all the
counter-flows which *must* be subtracted before comparison with the interior
transport calculated from the wind-stress curl using the Sverdrup relation
(9.21).  There is some evidence for a strong *sub-surface* counter-current offshore of the Gulf Stream.  This and other counter flows have not been well
observed.  Thus the discrepancy between the net western boundary region
transport and the Sverdrup transport may not be as large as presently thought
and further investigation is clearly needed.

# CHAPTER 12
# **Waves**

## INTRODUCTION

The word *waves* usually brings to mind a picture of undulations on the surface
of the sea or a lake, often with some semblance of regularity, and usually
progressing from a region of formation to a coast where they are generally
dissipated as surf or may, in part, be reflected. Less evident are the related movements of the water below the surface, waves which are entirely
beneath the air/water surface (internal waves) and a variety of waves which
are not usually evident by visual observations. The main classes of waves
and their causes are:

1. *ripples, wind waves* and *swell* - due to the effects of the wind on the
   air/water interface,

2. *internal waves* - which may occur when vertical density variations are
   present - various causes, e.g., current shear, surface disturbances,

3. *tsunamis* - generated by seismic disturbances of the sea bottom or shore,

4. *gyroscopic-gravity waves* - (surface and internal) of sufficiently long
   period that the Coriolis effect is important - various causes, e.g., wind
   stress changes, atmospheric pressure changes,

5. *Rossby* or *planetary waves* - large-scale and long period, evident as time-
   varying currents - various causes, e.g., time variations in wind stress
   and perhaps baroclinic and/or barotropic instability (mentioned in
   Chapter 11),

6. *tides* - due to fluctuating gravitational forces of the sun and moon.

In this chapter we will discuss some of the characteristics of the first three
classes above, describe the fourth and fifth classes briefly and leave the
discussion of tides to Chapter 13.

The classical approach to the study of waves is to consider the fluid dynamics
of ideal waves which, in side view, have a sinusoidal shape, and to progress
to other regular surface shapes. This approach gives a great deal of information about the relations between the surface shape, the progress of the waves
and the motions of the water below the surface. The least satisfactory
feature of this approach is that the ideal, regular waves thus studied bear
only a limited resemblance to real waves observed at sea which are character-
ised by their irregularity in form and period.

A more recent and pragmatic approach is to start from observations of the
shape of the irregular sea surface, regard it as a composite of a wide range
of possible ideal components and carry out spectral analysis to determine the

L - WAVELENGTH
T - PERIOD (TIME BETWEEN SUCCESSIVE CRESTS PASSING A FIXED POINT
C - SPEED (CELERITY) RELATIVE TO WATER    ( C = L/T )
H - WAVE HEIGHT  =  2 x AMPLITUDE (A)
h - DEPTH OF WATER BELOW MEAN SURFACE LEVEL

   N.B. In this figure, H is exaggerated relative to L for clarity.

Fig. 12.1.    Terms related to ideal (sine) waves.

characteristics of the spectrum of components, with particular emphasis on
the energy associated with the various components.  For example, if we have
observations of the surface elevation, η, at a point for a period of time, we
can consider this record to be the sum of sine waves of different amplitudes,
phases and frequencies.  Spectral analysis then consists of finding the
amplitudes and phases as a function of frequency.  A plot of amplitude squared
(proportional to the energy) *versus* frequency is called a *wave energy
spectrum*.  From the spectrum and the classical theory for each component we
should be able, in principle, to calculate the total effect of the wave field
by summing over all the components using the appropriate amplitudes and
phases.  To get a complete picture, the direction of travel should be considered as well and a spectrum including the direction of travel information
is called a *directional spectrum*.  This *statistical* approach is required in
applications of our knowledge of waves, e.g., to determine the total effects
of waves on ships and engineering structures near the sea surface.

Here we will start with the classical approach in which the wave shape is
assumed to be a sine wave.


SOME GENERAL CHARACTERISTICS OF WAVES

Assuming that the waves on the sea surface are simple sine waves (in vertical
section) some terms which we will use are illustrated in Fig. 12.1.   The
quantity (H), called the *height* of a wave (the vertical distance from trough
to crest), is twice what the physicist calls the 'amplitude' (A) of the
vertical oscillatory motion of the surface (the maximum displacement) above
or below the mean water level).

For all waves, the *speed* C = L/T where L is the *wavelength* (distance from crest to crest or trough to trough) and T is the *period*; the height H is basically independent of C, L or T.   (The symbol C comes from the alternative word 'celerity' used in the older literature.)

For convenience in referring to them, it is common to classify surface waves according to their periods as follows:

Table 12.1    Waves Classified by Period

| Period | Wavelength | Name |
|---|---|---|
| 0   - 0.2 s | centimetres | ripples |
| 0.2 - 9   s | to about 130 m | wind waves |
| 9   - 15  s | hundreds of metres | swell |
| 15  - 30  s | many hundreds of metres | long swell or forerunners |
| 0.5 min - hours | to thousands of km | long period waves including tsunamis |
| 12.5, 25 h etc. | thousands of km | tides. |

In all these surface waves gravity is the primary restoring force, allowing oscillations to occur.  (If some water is lifted up and allowed to fall back under the action of gravity its inertia will cause it to overshoot the equilibrium position; pressure forces will then push it back up and oscillations will occur.)   The ripples are also affected by surface tension; these waves are of very small amplitude and will not be discussed here.   For surface waves with periods of several hours, the Coriolis force must also be included in the analysis as will be discussed briefly near the end of the chapter.   Tides will be discussed in Chapter 13.

The ranges of periods of wind waves and swell actually overlap considerably - wind waves may have periods up to 15 seconds or so if the wind speed is very large, while swell with periods of only a few seconds is possible. *Wind waves* are the locally generated waves.   As they have a fairly wide range of directions the sea surface is quite irregular. *Swell* is the term for waves which have been generated elsewhere; it travels in one direction and is much more regular.   Also, as we shall show, the longer waves travel faster than shorter ones and so at some distance from the source area, at any one time, the swell has a narrow range of frequencies which also makes it more regular than wind waves.

SMALL AMPLITUDE WAVES

The word 'small' here is used in a comparative manner and refers to the 'relative height' or *steepness*, H/L.   For the simple theory to be correct within a few percent, this ratio should be less than about 1/20 and in many cases for real waves is 1/50 or less.   (For clarity in Fig. 12.1 and other figures we have exaggerated the wave height in relation to the wavelength.)   Here we shall usually consider the first order theory, i.e., we neglect terms which are of order H/L (or higher powers, e.g., $H^2/L^2$) times the terms retained.   (The extra terms arise from the non-linear terms in the equation of motion but the non-linear ones remain small compared with the other terms, i.e., are of higher order in H/L.)

Fig. 12.2.    Properties of deep and shallow water waves.


For a progressive sine wave, the displacement η of the free surface from the
mean level is given in terms of time t and displacement in the x direction
(for a wave travelling in the x direction) by:

$$\eta \;=\; A \cdot \sin 2\pi \left( \frac{t}{T} - \frac{x}{L} \right) . \qquad\qquad (12.1)$$

For such waves it can be shown that the speed:

$$C \;=\; \sqrt{ \left( \frac{g \cdot L}{2\pi} \cdot \tanh \frac{2\pi h}{L} \right) } \qquad\qquad (12.2)$$

where g = acceleration due to gravity, 'tanh' is the hyperbolic tangent and
h is the water depth.  For $h/L > 1/2$, tanh $2\pi h/L = 1$ within 0.5% while for
$h/L < 1/20$, tanh $2\pi h/L = 2\pi h/L$ within 3%, so that the expression for C may be
simplified as followes:

(1)   for h > L/2 , called DEEP    water waves, then $C_d = \sqrt{g \cdot L/2\pi}$ ,
                                                                              (12.3)
(2)   for h < L/20, called SHALLOW water waves, then $C_s = \sqrt{g \cdot h}$  .

Some authors refer to deep-water waves as 'short' waves, i.e., L is short
compared with h, and to shallow-water waves as 'long' waves, i.e., L is long
compared with h.   Fig. 12.2 illustrates these nomenclatures.

Fig. 12.3 shows plots of equation 12.2 as speed C against water depth h for a
selection of wavelengths from 10 m to 10 km.   The left-hand (straight) line is
the plot of $C_s = \sqrt{g \cdot h}$  (shallow-water wave speed).   Then the line for
L = 200 m (for example) shows that the speed follows the shallow-water line
up to about 10 m water depth (h = L/20) where it commences to curve to lower
values, eventually reaching its constant value of $C_d = 17.7\,\mathrm{m\,s^{-1}}$ at about
100 m water depth (h = L/2).   The zone on the graph to the right of the dashed
line is where the deep-water speed approximation holds, and the intermediate
zone between the shallow-water speed line and the dashed line is where the

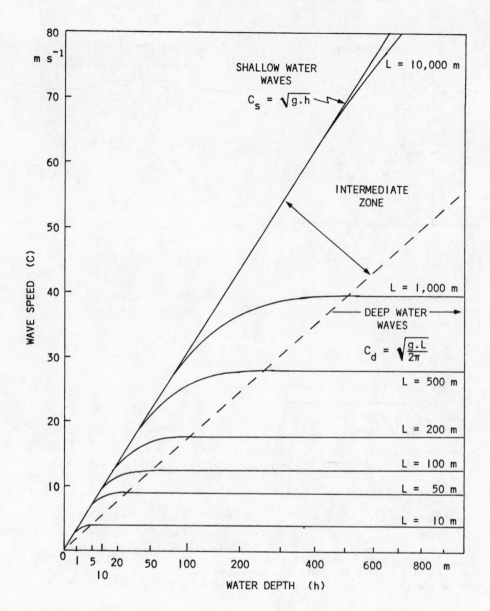

Fig. 12.3.    Wave speed versus water depth for various wavelengths.

full expression of equation 12.2 must be used to calculate the speed.   In
practice, the shallow and deep water approximations find most use, the intermediate zone applies chiefly in studies of the surf zone.

If we introduce the values for the constants in the two expressions for the
wave speed we obtain the expressions in Table 12.2 in which are also included
a few numerical values for wave properties.

Table 12.2  Deep and Shallow Water Wave Formulae and Sample Values

DEEP                                          SHALLOW

$$C_d = \sqrt{g \cdot L_d / 2\pi} \qquad\qquad C_s = \sqrt{g \cdot h}$$

$$C_d = 1.56\ T = 1.25\ \sqrt{L_d} \qquad\qquad = 3.13\ \sqrt{h}$$

$$L_d = 1.56\ T^2$$

(L & h in m, T in s, C in m s$^{-1}$, g = 9.8 m s$^{-2}$)

Examples

| | | | | | |
|---|---|---|---|---|---|
| T = | 5 | 15 s | h = 5 | 20 | 4,000 m |
| | wind | swell | | | (tsunami) |
| | wave | | | | |
| $C_d$ = | 7.8 | 23 m s$^{-1}$ | $C_s$ = 7 | 14 | 200 m s$^{-1}$ |
| or  = | 28 | 84 km h$^{-1}$ | = 25 | 50 | 710 km h$^{-1}$ |
| $L_d$ = | 39 | 350 m | | if $L_s$ ≈ | 200 km |
| | | | | then T ≈ | 17 min. |

The numerical values for the deep-water waves give an idea of the properties
of wind waves and swell, while the first two examples for shallow-water waves
show the retarding effect of shoaling water on such waves.  The last example,
for h = 4,000 m, may seem out of line for *shallow* water, but it is included
to emphasize that the term 'shallow' is only relative (see Figs. 12.2 and
12.3).  The example is typical for the (quite long) tsunami waves generated
by underwater seismic disturbances.

Another point to notice is that the speed of deep-water waves depends on their
wavelength and so on their period, i.e., they are dispersive waves.  This term
refers to separation in speed along their direction of travel, not to separation in direction, although it also occurs.  The speed of the longer deep-
water waves is greater than that of the shorter ones.  Therefore, if a number
of waves of different wavelengths (a spectrum of wavelengths) are generated
simultaneously, the longer ones will move ahead of the shorter ones and be
observed first at a distant point (hence the term 'forerunners' for the
longer period, i.e., longer wavelength, waves generated by the wind).  Also,
shorter waves tend to lose their energy by frictional effects somewhat faster
and die out sooner than longer ones, and so do not travel so far.

A consequence of this dispersion is that by observing the swell for a few
days at one location it is often possible to determine how far away was the
storm which generated the waves.  If the spectrum of swell periods is determined at intervals of a few hours at the wave recording station (in relatively

Fig. 12.4.  Orbits of water particle motion for deep and shallow water waves.

deep water) it will be observed that the mean period decreases steadily with time.  Because the speed of travel of deep-water waves is proportional to their period (Table 12.2), the difference between the times of arrival of the longer period (early arrival) and the shorter period (late arrival) swell is due to their different speeds.  It is then possible to calculate from what distance the swell must have come to yield the observed time separation at the wave station.  The reader should note that in practice the wave records will provide information about mean periods for groups of waves, not waves of a single period, and the calculation is a little more complicated than might appear at first sight.  For a group of waves it is necessary to use the 'group speed', $C_g$, not the 'phase speed'.  For deep-water waves, it can be shown that the group speed is one-half of the phase speed ($C_g = C_d/2$), while for shallow-water waves it is the same as the phase speed ($C_g = C_s$).

The observations give only the distance to the point of generation, not the direction, but if observations are available at two separate wave stations for the same generation event, then the intersection of the two radial distances from the stations will indicate the location of generation.

Orbital Motion of the Water Particles

It is only the shape of the wave which moves forward at the speed $C_d$ or $C_s$; the water particles themselves do not travel across the ocean but rotate in _orbits_, circular for deep-water waves and elliptical for shallow-water waves. These orbits decrease in size with increase in depth (Fig. 12.4).  For deep-water waves the diameter of the orbit is $D_z = H \cdot \exp(2\pi z/L)$ where H is the wave height at the surface and z is the level (numerically negative with z = 0 at the average surface level as usual).  For example, at z = - L the orbit diameter will be only 0.002 of that at the surface (Fig. 12.4a).  For shallow-water waves the orbits are already elliptical at the surface (Fig. 12.4b); the horizontal dimension decreases only slightly while the vertical dimension decreases markedly with increasing depth, until at the bottom (if it is flat) the motion will simply be back and forth.

Fig. 12.5.   Refraction of waves approaching a smoothly shelving beach.

If we consider higher order corrections the orbits are not quite closed; in deep water there is a net flow, in the direction of travel of the wave, of magnitude $(\pi^2 \cdot H^2/L^2) \cdot C_d \cdot \exp(4\pi \cdot z/L)$. This net transport is called the *Stokes drift*. For L = 100 m, H = 3 m, then $C_d$ = 12.5 m s$^{-1}$ while the Stokes drift at the surface is only 0.1 m s$^{-1}$. The speed of the orbital motion in deep water is $(\pi \cdot H/L) \cdot C_d \cdot \exp(2\pi \cdot z/L)$, so the net flow is only a small fraction $[(\pi \cdot H/L) \cdot \exp(2\pi \cdot z/L)]$ of the orbital speed. In shallow water there is also a Stokes drift and additional effects due to bottom friction. In either case, as the wave reaches breaking conditions the Stokes drift may be several percent of the phase speed. In the surf zone, the net onshore transport must, by continuity, be balanced by offshore transport, which often takes the form of narrow jets to seaward, called *rip currents*, which individually last for only a few minutes. Such rip currents are often too fast to swim against toward the shore – the best tactic if caught in one is to swim either way parallel to the shore to get out of the usually narrow outward current.

If the waves are approaching the shore at an angle (outside the refraction zone, see next section), the net transport may have a component along the shore giving a longshore current. It will be slow but may be significant over a long period in transporting sand, etc., along beaches after the material has been stirred up by the waves.

### Refraction and Breaking in Shallow Water; Diffraction

Shallow water waves all travel at the same speed $C_s = \sqrt{g \cdot h}$ in water of a given depth h, and therefore do not show dispersion of speed, but where the bottom depth is changing their direction of travel may change. More generally, as waves move into shallow water their period remains constant but C decreases and therefore L decreases. As an example, Table 12.3 shows the decrease of

Fig. 12.6.  Wave refraction on approaching an underwater ridge (left) and a
            valley (centre).  Dashed lines represent depth contours, full
            lines represent wave orthogonals (directions of travel of waves).


speed and wavelength for waves of period 8 seconds on entering shoaling
water.

Table 12.3  Decrease of Speed and Wavelength in Shoaling Water for Waves
            of Period 8 s and Length 100 m in Deep Water

| | | | | | |
|---|---|---|---|---|---|
| h = | 50+ | 10 | 5 | 2 | m |
| C = | 12.5 | 8.9 | 6.6 | 4.3 | m s$^{-1}$ |
| L = | 100 | 71 | 53 | 35 | m |

Hence, if a series of parallel crested waves approaches at an angle to a
straight shoreline (Fig. 12.5) over a smooth sea bottom which shoals gradually, they progressively change direction as the end of the wave nearer the
shore (A in Fig. 12.5) slows down earlier than that farther away (B in Fig.
12.5).  As a result, the waves become nearly parallel to the shore by the
time that they pile up as surf.  The change of direction associated with
change of speed is called *refraction*.  The same phenomenon occurs abruptly
to light waves travelling from air to water but gradually, as for water
waves in this case, for light coming from the sun and entering the upper
atmosphere at an angle to the vertical.

(a) DEEP WATER
    H/L SMALL

(b) SHOALING WATER
    H/L INCREASES

(c) NEAR BEACH
    H/L → 1/12

SURF
ZONE

BOTTOM

(N.B. H/L exaggerated above)

Fig. 12.7.    Shape of waves: (a) in deep water, (b) in shoaling water,
              (c) close to beach.


If the sea bottom does not have a uniform slope along the full length of the
shore, the refraction may be more complicated.  Two simple examples are where
there is an underwater ridge running out at right angles to the shore, and an
underwater valley.  The refraction pattern for waves coming straight in from
offshore would then be as in Fig. 12.6.  In this figure are shown not the
wave crests but the wave *orthogonals* which indicate the direction of travel
of the waves, as do the arrows in Fig. 12.5.

Refraction of waves round a headland, for instance, occurs if the water
deepens gradually to seaward from the land, but not if the water is of relatively uniform depth off the headland.  Waves are often observed to be refracted round islands and one can sometimes see an interference pattern set
up where the waves which are refracted around the two sides of a small island
meet behind it.

As the waves move inshore and slow down, not only does the wavelength decrease
but also the wave height changes.  It can be shown that the product of the
wave energy per square metre times the group speed ($C_g$) is constant as the
waves move inshore (until they break), with a value averaged between two
successive crests of $(\rho \cdot g \cdot H^2 \cdot C_g)/8$ Joules per metre width of crest per
second.  If the waves are initially long (e.g., tsunamis) $C_g = C_s$, both
decrease and H increases.  However, if the waves are initially short, at
first $C_g$ increases, reaching a maximum value of 1.2 times the deep-water value
when $h/L \simeq 0.19$ where L is the local value, not the deep-water value .  In
this zone, H decreases to a minimum of about 90% of the deep-water value when
$C_g$ is a maximum.  H/L is nearly constant at first but begins to increase
before $h/L \simeq 0.19$ because L decreases faster than H.  At h/L = 0.19, the
steepness H/L is about 10% greater than the deep-water value.  As $C_g$ decreases
as the waves move further inshore, H must increase.  However, the decrease

in L dominates H/L changes.  In the example given in Table 12.3 when h = 2 m,
L is about 35% of the deep-water value but the increase in H is only about
25%.  For initially long waves, L decreases proportionally to $C_s = \sqrt{g \cdot h}$
while H increases proportionally to $(C_s)^{-\frac{1}{2}}$.  There is a limit to how much
H/L may increase.  Theory puts this limit at H/L = 1/7 but in practice it is
rare for waves to get steeper than H/L ≈ 1/12.  When the wave steepness
approaches this limit the waves tend to change from a symmetrical sine shape
(Fig. 12.7a) to a more peaked shape (Fig. 12.7b) and finally as H/L reaches
about 1/12 the waves become unstable and break as surf (Fig. 12.7c).  In the
very shallow water where the ratio H/L is approaching the practical limit of
about 1/12, the waves tend to behave as individuals, rather than as a successive train, and they become progressively more unsymmetrical in side view
until they break.  Usually, rather shallow water must be reached; breaking
does not occur until the wave height is comparable to the water depth or until H/L ≈ 1/12, whichever occurs first.

*Diffraction* can also occur with water waves, as for light waves.  For example,
if waves arrive at a harbour entrance, some of their energy will continue
into the harbour in the original direction of wave travel, but near the sides
of the opening, some wave energy will be diffracted into the geometrical
'shadow' area behind the harbour walls.  This phenomenon is most conspicous
when the gap in a sea wall or reef is narrow compared with the wavelength of
the waves.  Then most of the energy entering the harbour is that which has
been diffracted and if the harbour has a fairly uniform depth, the pattern of
diffracted waves will be in the form of circular arcs centred on the gap.

## THE GENERATION OF WAVES

Wind waves are started by a wind blowing for some hours *duration* over a sea
surface many miles long called a *fetch*.  The fitful gusts of wind generate a
choppy and irregular *sea*.  These oscillations, once set up, continue to run
across the surface of the sea far beyond the direct influence of the wind.
Under these conditions they are called *swell*.  Swell consists of uniform wave
trains with a broad sideways extent of the individual crests.  Because it is
comparatively uniform we can numerically describe the height and period of a
wave at the beginning of the swell zone (i.e., at the end of the fetch) and
during its subsequent progress.  The swell *decays* for a long distance while
its wavelength increases and wave height decreases.  As the swell enters
shallow water it *feels bottom* and a rejuvenation takes place.  The wave speed
and length decrease and the height increases, but the period remains constant.
The swell finally peaks up into *waves*, breaks, and is dissipated as *surf*.

The above paragraph gives a brief sketch of the generation and dissipation of
wind waves.  Clearly, the wind is responsible for the generation of surface
waves which are almost always present.  How are we to get more quantitative
information on what sort of waves will be present under given conditions of
fetch, duration and wind speed?  Later we shall describe an empirical
approach, based on observations of the waves themselves under a variety of
conditions, which still has to be used for practical purposes because a
quantitatively accurate theory based on physical laws has yet to be produced.
Here we shall describe some of the steps that have been taken toward a
satisfactory theory.  (A more extensive discussion can be found in LeBlond
and Mysak, 1978, given in the Further Reading list.)

First, consider a turbulent wind flowing over a solid surface.    Somewhat
above the surface, the stress (or downward momentum flux) is due to the turbu-
lent Reynolds stresses ($-\rho \cdot \overline{u'w'}$ and $-\rho \cdot \overline{v'w'}$ as shown in Chapter 7).    As the
surface is approached, the amplitudes of the turbulent velocity fluctuations
are suppressed by the boundary condition of no flow parallel or perpendicular
to the surface.   At the same time, the mean shears ($\partial u/\partial z$ and $\partial v/\partial z$) increase.
If the surface is smooth (e.g., a sheet of glass) the final stress transfer
at the surface is by molecular viscous stresses ($\mu \cdot \partial u/\partial z$ and $\mu \cdot \partial v/\partial z$ in the
x and y directions, respectively).   If the surface is rough (e.g., a sand
beach) the bumps on the surface will cause the air flow to separate from them
leaving stagnant regions behind them.   There will be a positive pressure
difference from the upwind to the downwind sides of the 'roughness' elements.
Part, and perhaps most, of the final stress transfer will be due to these
pressure differences or what is termed 'form drag'.   As pressure is a normal
stress (force/unit area perpendicular rather than parallel to a surface) one
may also say that the final momentum transfer (flux) is mainly by normal
rather than shear stresses for a rough surface.

Now surface waves are nearly irrotational (i.e., have almost zero vorticity).
Irrotational motions are produced by normal stresses while rotational motions
are produced by shear stresses.   Thus, most of the wave generation must be due
to normal stresses (pressure).   The facts that waves are nearly irrotational
and nearly linear (the advective acceleration terms are of higher order in
H/L than the local acceleration terms) are important in explaining why waves
outside active generation areas, where wave breaking is important, decay very
slowly.   Thus, swell is not likely to produce turbulence and hence turbulent
friction to damp it.   Further, for exactly irrotational motion the molecular
viscous terms vanish identically.   Swell generated near Antarctica has been
traced across the whole Pacific to Alaska, albeit with considerable loss of
amplitude over this very long distance.   The decrease in amplitude may be due
to spreading of the energy because of differences in direction of travel as
well as to viscous losses.

One of the fairly early theories for wave generation, once the waves already
existed, was that they grew by form drag with flow separation at their crests.
This theory was proposed by Jeffreys in the mid-1920's.   Tests with flows over
solid models showed that the effect was too small to explain observed rates of
wave growth.   A problem with these tests, not clearly recognized at the time,
is that results for solid surfaces cannot be applied directly to a moving
fluid surface such as that of the ocean.   Another problem, not recognized for
a long time, is that a condition for flow separation is that wave breaking
must occur, which may be of fundamental importance as we shall see.

Renewed attempts to solve the problem were made in the late 1950's and early
1960's.   First, Phillips suggested a means for getting waves started on an
initially undisturbed surface.  He points out that as the air flow is turbulent, not only are there velocity fluctuations but pressure fluctuations as
well.   These pressure fluctuations may start wave motion; they lead to a
growth of wave energy proportional to time, t.   (There is some observational
evidence to support Phillips' theory of initial growth.)  Once the waves
exist they may modify the air flow so that the growth rate becomes proportional to the wave amplitude (or energy) and hence exponential in time.   Assuming
that waves of *small* amplitude have been formed, the growth process may be
calculated using linearized stability theory as was done by Miles.   However,
his calculated growth rate, though exponential, was soon found to be much
smaller than observations indicated.    The observations themselves show

considerable variation but there is little doubt that there is a real discrepancy.

Further ideas have been suggested in attempts to overcome this disagreement
between observations and theory, e.g., that the momentum input to the waves is
largely to the very short waves and ripples and that these in turn become very
steep and perhaps break (a very non-linear process) and transfer at least some
of their momentum and energy to the larger waves.  Again, while this mechanism
may play a role, it does not resolve the observational-theoretical discrepancy
either.

Recently, numerical calculations have been made showing the air flow over
water waves in more detail.  The rate of growth due to pressure depends quite
strongly on the wave steepness H/L.  These calculations also show that about
one-half of the total momentum transfer is through the pressure field (normal
stresses).  These results are supported by observational data as well as by
the argument that, to explain observed growth rates, much of the total stress
must go into the waves by normal stresses to produce nearly irrotational wave
motions.  The shear stress is also greatest near the crest, giving some
support to the idea of initial input to short waves which then transfer their
momentum to longer waves.  However, even these results do not lead to growth
rates large enough to explain the observed values.

Other very recent observations have suggested that the momentum transfer is
much increased when wave breaking occurs.  Thus, Jeffreys' original argument
may be reasonably correct *if* breaking is taken into account (which Jeffreys
did not do explicitly).  Non-linear transfer from shorter to longer waves no
doubt plays a role too.  A quantitatively correct wave generation theory
remains to be established but the recent results suggest further research
which may lead to it.

Other approaches are being taken too.  If one observes the growth of the wave
spectrum and calculates the non-linear transfer and viscous dissipation, the
input function can be calculated by differencing.  Attempts at the direct
measurement of the input have been and continue to be made.  Such measurements
are difficult as the reader may appreciate if he considers how to make observations in a breaking-wave field while trying to keep instruments *at* the
moving surface.  Of course, neither sort of observation 'explains' the wave
generation process but they do give the theoretician something to try to
reproduce.

Non-linear effects in surface waves are weak but they are not negligible.  In
a fully developed sea (one for which fetch and duration are not limiting, and
in which further growth no longer occurs because wave breaking balances the
input from the wind) there are components in the spectrum whose phase speeds
are greater than the wind speed a few metres from the surface.  Indeed, the
peak values of the spectrum have this characteristic.  If one translates to
coordinates moving at the phase speed of the waves near the spectral peak then
the air flow is contrary to them and it is very difficult to imagine how the
wind can enhance them.  However, this argument may not be correct.  In reality,
one always deals with groups of waves.  In coordinates moving at the group
speed of the waves near the spectral peak, the air flow is still such that it
may do work on the waves and enhance them.  (In Phillips' original study of
the initiation of waves he considered the phase speed and got a growth rate
proportional to $t^2$, but because the waves travel in groups at the group speed
they are always falling behind or getting out of phase with the pressure

fluctuations, and the growth rate is really proportional to t.)  The calculations on non-linear transfers also indicate that the longer waves in the
spectrum gain energy from non-linear transfer from shorter waves.

Considering that much of the momentum input at the surface goes, at least
initially, into the waves one may wonder about using the total stress as a
forcing function for large-scale ocean circulations.  What if the waves
radiate away and take their momentum somewhere else or even to a distant
shore?  It seems that this possibility is not serious.  The wave field develops quite rapidly; in the developed stage the momentum input to the waves is
transferred through wave breaking to the current quite quickly (with transfer
to longer waves a possible intermediate stage).  Only in rather small regions
of rather strong winds is significant radiation of momentum out of the area
likely; even then on a global scale most of the momentum goes into the
currents locally.  Thus, although much of the momentum input from the wind
passes through the wave field, the net input to the wind-driven circulation is
probably not affected significantly.

Finally, a comment should be made on the constancy or near constancy of the
drag coefficient.  For a solid rough surface the drag coefficient (defined in
Chapter 9) is constant (provided that increased wind speed does not change
the geometry, e.g., a hay field becomes flatter and 'smoother' as the wind
becomes stronger, but a field of boulders does not).  The drag coefficient
over the ocean is constant or perhaps slowly increasing as wind speed increases.  Thus, to the air flow the sea presents a constant (or nearly constant) roughness.  (An observer riding on a ship would hardly agree with the
air flow on this point!)  Perhaps the apparent roughness is due to the short
waves which develop quite quickly; perhaps the effect(s) of flow separation
occurring over longer distances with longer waves at higher winds gives an
equal effect.  Further investigations are clearly in order.  There is some
evidence that the drag coefficient is higher when the wave steepness is
greater although the scatter in values is so large that disagreements about
this interpretation are possible.  Again, further investigation is needed.

## MEASUREMENT OF WAVES

A number of methods are available for obtaining information on wave heights
and periods, some approximate and some accurate.

The simplest way of all is simply to look at the sea and make a visual estimate.  It takes a lot of practice with comparisons with actual measurements to
obtain reliable data in this way.  The next simplest way is to make visual
observations of the water surface against a vertical scale mounted on a pier
in shallow water or, in deep water, on a float with a deep horizontal plate
'damper' to limit vertical movements of the float and scale.  Because the
period of wind waves is relatively short, only a few seconds, such visual
measurements are limited to estimates of the height of only a proportion of
the waves.  Because of the variety of heights present in most real wave conditions it is common to quote the mean height of the highest one-third of the
waves (called the *significant wave height*, $H_s$) as a descriptive characteristic.
For many purposes, although not for all, this value is more useful than the
value for the single highest wave, for example.

A third method is to use a fixed pressure-sensor mounted below the depth of
the deepest troughs.  The hydrostatic pressure below surface waves varies

periodically as the depth of water from the surface to the sensor varies and
so a continuous record of pressure against time will give information on the
surface shape.  This method is only practical when the sensor can be mounted
in a fixed position at a relatively small distance below the troughs because
for deep-water waves the pressure variations decrease rapidly with depth in
the same manner as the orbit radius, so that the deeper the sensor, the
smaller are the pressure variations available to measure.  For shallow water
waves, a sensor mounted on the bottom will record almost the full pressure
variations due to the variation of the height of the water column as the waves
pass over the instrument, so that pressure sensors can be used effectively
near shore.  They can also be used even in the deep ocean (thousands of metres)
to measure tsunamis because these are shallow-water waves.

Information about the details of the surface height variations when waves are
present has been obtained with electrical devices mounted to penetrate the
sea surface.  One method is to mount on a vertical rod, at intervals of a few
centimteres, a series of pairs of wires with a small gap between the wires of
each pair.  Then the wire pairs which are immersed in the conducting sea water
will be short-circuited, and by recording continuously against time the number
of pairs from the bottom which are short-circuited a record of sea surface
level is obtained.  An alternative method is to use a vertical, bare resistance
wire and then a record against time of the value of the resistance from the
top of the wire to the sea will yield a record of the height variations of the
sea surface.  A third alternative is to mount a thin, insulated wire vertically through the sea surface and to measure the electrical capacitance between
the wire as one electrode and the sea as the other.  Of course, there are
technical difficulties to be overcome.  For the first method, drops of water
may remain between the wires when the surface falls below them, or in the
other two methods a thin film of water may be left behind briefly as the water
level falls, so indicating too great a water height.  (Remember that we would
be dealing with fluctuations of water level of only seconds duration.)  The
use of a hydrophobic wire for the capacitance wire system helps to minimise
this source of error.

A method developed by the National Institute of Oceanography (now the Institute
of Oceanographic Sciences) in England makes it possible to measure waves at
sea from a ship.  The water pressure is measured at a low point on the hull,
normally below even the troughs of the waves, to give the wave height relative
to the ship.  At the same time, the vertical motion of the ship is measured
with an accelerometer (integrated twice to show vertical height variations)
and added algebraically to the pressure record.  Wave profiles may also be
obtained from the records of a vertical accelerometer mounted on a buoy floating on the surface of the sea.

All of the above methods have a major failing - they provide information at
one point only.  If real surface waves were a sum of pure sine waves travelling
in a single direction the problem would not be serious, but even a few minutes
observation will show that the real water surface is usually quite irregular,
small waves superimposed on larger waves superimposed on swell, and the crests
of the waves generally are quite short (only a few wavelengths at most) so
that the real water surface varies with both x and y and with time.  To obtain
more complete information, stereo-photographs of the sea surface may be made.
However, the analysis of such photographs is a very laborious process and this
method has not been used much.

Fig. 12.8.    Character of real wave shape (profile) to compare with ideal
              sine waves of Figs. 12.1 and 12.2.  Note that the vertical scale
              is much exaggerated as before.


Another method to obtain information on the spatial structure has been to fly
over the sea at as nearly constant altitude as possible (measuring and
correcting for variations) and record the shape of the sea surface with a
narrow-beam radar altimeter, or more recently, a pulsed laser altimeter (to
look at only a very small area of the sea surface at a time).  This method
only yields the surface elevation along the flight path but flights can be
made in several directions and a statistical picture built up of the sea
surface shape.

Finally, to test some aspects of wave theory it is desirable to obtain information on the direction of propagation of waves.  In principle this information can be obtained by mounting a number of wave recorders in a geometrical
pattern and examining the phase relationships between the records.  The
results are limited by the number of measuring points and the analysis is complicated.  In principle, this directional information can also be obtained
from the stereo-photographs or laser altimeter records.

Real Waves

The records from such instruments and procedures make it clear that the sea
surface rarely has the ideal sine shape as in Fig. 12.1 and 12.2 except for
swell, but is more likely to look like Fig. 12.8 because there are usually a
large number of wave components present simultaneously.  The only practical
way to deal with this situation is the statistical one mentioned earlier in
which the spectrum of wave energy is related to wave period, as in Fig. 12.9
(to be discussed shortly).


## WAVE GENERATION BY THE WIND; SEMI-EMPIRICAL RELATIONS

Although there is still uncertainty in our knowledge of many details of the
actual mechanisms of wave generation at the sea surface by the wind, many
observational data on related wind and wave properties have been accumulated
and graphical relations assembled.  Some features of one of the sets of relations (Pierson, Neumann & James, 1955) will be described to illustrate their
character.  The wind factors are *wind speed, fetch* (the linear distance over
which the wind is blowing over the sea) and the *duration* (the time for which
the wind has been blowing over that fetch).  Wave properties are the *significant wave height* $H_s$ (the average height of the one-third highest waves) and
the range of *wave periods* or *frequencies* in the wave spectrum.

Fig. 12.9.    Idealized spectra of wind-wave energy versus frequency and
              period for three wind speeds for a fully developed sea.


Fig. 12.9 shows three plots of the square of wave height ($H^2$) against frequency, for a wave system developed on the sea by winds of 10, 15 and $20\,m\,s^{-1}$,
respectively.  One feature of these curves is that the quantity $H^2$, which is
related to wave energy, increases very much more rapidly than wind speed;
another feature is that the spectrum of wave energy as a function of frequency
is peaked and that the peak occurs at lower frequencies (longer periods) at
higher wind speeds.  These curves are for a *fully developed sea*, i.e., when
the wind has been blowing for long enough and with a sufficient fetch for the
steady state to be established with the energy spectrum at a maximum for that
speed.  The numerical information on which such spectra are based is obtained
from measured wave records at sea at various wind speeds.

From such energy spectrum curves are developed *co-cumulative spectrum curves*
such as those in Fig. 12.10 (full lines for wind speeds of 10, 15 and $20\,m\,s^{-1}$).
The ordinate for any point on each curve is proportional to the total cumulative wave energy from infinite frequency (zero period) to the frequency represented by the point on the curve.  (Note that the frequency scale increases to
the right but the period scale increases to the left.)  In these plots the
ordinate (energy scale) has been arranged so that it is linear in significant
wave height while the abscissa is linear in period from zero to 20 seconds
and then compressed for longer periods.

Fig. 12.10.   Co-cumulative wave spectra as significant wave height (H$_s$) and
              wave energy against frequency (f) and period (T) for three wind
              speeds (full lines), four fetches (dotted lines) and four dura-
              tions (dashed lines) (adapted from W.J. Pierson, G. Neumann &
              R.W. James, *U.S.N.H.O. Publ. 603*, pp. 68, 69, 1955).

To illustrate the information available in this plot we will take the 15 m s$^{-1}$
wind-speed curve (full line) as an example.  Reading from the right, the
curve indicates that the cumulative wave energy (and the significant wave
height H$_s$) increases with period slowly at first, then more rapidly, then more
slowly and finally levels out (expressed as H$_s$ at about H$_s$ = 6.3 m) a little
before 20 s period.  The fact that there is a maximum value indicates that
for any given wind speed there is a maximum total energy possible (wave breaking causing the limit). This maximum is seen to increase with increased wind
speed.  The steepest part of the curve (at about 12 s period for the 15 m s$^{-1}$
wind speed) corresponds to the peak of the H$^2$ spectrum curve of Fig. 12.9.

In Fig. 12.10, in addition to the three sample wind-speed curves (full lines),
there are two sets of cross lines, dashed lines for 10, 20, 30 and 40 hours
duration (assuming unlimited fetch) and dotted lines for 100, 500, 1,000 and
1,500 km fetch (assuming unlimited duration).  These lines indicate the pro-

gressive development of wave height and period with these parameters.  For instance, the 20 h duration line intersects the $15\,\mathrm{m\,s^{-1}}$ wind speed line at 5.8 m significant wave height and 14 s period.  The former indicates that this value of $H_S$ is reached after 20 h and the latter indicates that the majority of the wave energy will be present at periods of 14 s *or less*.  Alternatively, in terms of fetch the (dotted) 500 km line indicates that at the end of this distance, for a steady wind of $15\,\mathrm{m\,s^{-1}}$, the value of $H_S$ will be 6.2 m and most of the wave energy will be at periods of 17 s or less.

The levelling off of the wind speed lines indicates that the sea is fully developed, i.e., that the rate of input of energy and momentum by the wind is balanced by the rate of loss due to wave breaking.  In principle, this fully developed state requires infinite duration and fetch but in practice it is considered to be effectively reached in a finite duration and fetch as shown in Table 12.4:

Table 12.4  Effective Duration and Fetch for a Fully Developed Sea, With Corresponding Significant Wave  Height and Range of Wave Periods

| Wind speed | $(\mathrm{m\,s^{-1}})$ | 5 | 10 | 15 | 20 | 25 |
|---|---|---|---|---|---|---|
| Duration | (h) | 2.3 | 9.5 | 22 | 40 | 64 |
| Fetch | (km) | 20 | 130 | 480 | 1200 | 2400 |
| $H_S$ | (m) | 0.4 | 2.2 | 6.1 | 13 | 22 |
| Range of periods (s) | $T_U$ | 6 | 11 | 16 | 21 | 26 |
|  | $T_L$ | 1 | 3 | 5 | 6 | 8 |

The wave periods in Table 12.4 are defined such that 5% of the total energy will be at periods greater than $T_U$ and 3% at periods less than $T_L$, i.e., 92% will be between these two periods.  For a non-fully developed sea, either the duration or fetch may be the limiting factor; which should be used will be determined by the conditions for which the wave calculations are to be made.

Statistical studies indicate that for a long series of waves the average height of all waves, the significant wave height (highest one-third) and the average of the one-tenth highest waves will be in the ratios 0.6:1.0:1.3.  It must also be realized that there will be a range of heights in the highest one-third waves and it is to be expected that the longer the series of waves observed, the higher will be the highest ones.  For instance, for 100 waves observed there is a 1/20 chance that the highest will be over 1.9 $H_S$, while for 1,000 waves there is a 1/20 chance that the highest will be over 2.2 $H_S$.

The above examples have been drawn from the Pierson, Neumann & James procedure. It should be pointed out that other investigators have also analysed wave observations and developed wind/wave relationship graphs and calculation procedures and that all methods do not give identical results.  The differences may be due to differences in the wave characteristics in the different regions from which the data were drawn, to differences in observing techniques or to differences in the treatment of data.  However, the procedure described above gives an indication of how an empirical approach may be employed to obtain useful results even though the generation mechanism is not understood in detail.

There are numerous applications for wind/wave relationships.  One of the needs
which led to the development of such procedures (by Sverdrup and Munk in the
first instance) was to forecast wave conditions, from forecast winds, for
beach landings during World War II.  The forecasting of wave conditions for
other operations, such as for laying undersea cables and pipes is a more
recent application.  The calculation of wave characteristics for regions for
which wind data are available, but for which no wave data are available, can
be important for ship hull design or for ship routing (because ship speeds are
decreased by increased wave heights).  For structures such as oil drilling
rigs at sea, an important prediction from wave statistics is the probable
highest wave over a period of, say, 10 or 100 years to avoid going to the
often considerable extra expense of building the structure stronger than it
need be.

## ENERGY OF WAVES

It is obvious that a water surface with waves will have more potential energy
than when the surface is level, because one can consider that the water in the
troughs  has been raised vertically to form the crests.  Also, as the water
particles are moving, there must be kinetic energy associated with waves.  It
can be shown that for sinusoidal waves, when averaged over a whole wave, the
average energy is $E = \rho \cdot g \cdot H^2/8$ Joules $m^{-2}$, of which one-half is potential
and one-half kinetic energy.  Notice that this expression is independent of
wave length or period.

It can also be shown that the wave energy travels at the group speed which, in
deep water, is equal to one-half the phase (wave) speed (i.e., in deep water,
wave energy travels at $C_d/2$).  In shallow water the energy travels at the
phase speed ($C_s$).

R.S. Arthur estimated that the surface wave energy of the world oceans at any
time is about $10^{18}$J and that the rate at which energy reaches the west coast
of the United States (wave power) is about $4 \times 10^{10}$ J s$^{-1}$ = $4 \times 10^7$ kiloWatts
($5 \times 10^7$ horsepower) or about 25 times the power generated by a fairly large
hydroelectric station.  The power reaching the total shoreline of all the
oceans was estimated at about $2 \times 10^9$ kW.  If all this energy were converted
to heat and distributed without loss uniformly through the oceans it would
take about 90,000 years to raise their temperature by $1 C°$, i.e., the rate
of heat contribution to the oceans by dissipation of wave energy is negligible
compared with the solar contribution (about $3 \times 10^{12}$ kW).  The rate will be
higher in the surf zone where the wave energy is dissipated but even there it
will be small and likely to be removed by circulation too quickly to be
detected; measurements have not shown any significant temperature rise in the
surf zone.

## TSUNAMIS OR SEISMIC SEA WAVES

*Tsunami* is the transliteration of a Japanese word meaning 'harbour wave' (as
distinct from the ordinary tidal rise and fall), and it is now generally used
to refer to long water-waves generated by sea bottom movements associated
with earthquakes.  The term 'seismic sea wave' is also used.  The older term
'tidal wave' is incorrect and should not be used because the generating
mechanism for tsunamis is quite different from that for the normal tides.  It
is known that virtually all tsunamis follow earthquakes occurring under the

sea or near the shore, but not all earthquakes generate tsunamis. It is considered probable that only those earthquakes which involve a significant
component of motion perpendicular to the bottom, i.e., a rise, fall or tilt,
are likely to be generators, while those in which the motion is horizontal do
not generate tsunamis.

Occasionally, tsunamis are generated by other earth movements, such as large
landslides into the sea, or are possibly associated with marine volcanic
activity, but the effects are usually not wide-spread, whereas tsunamis
generated by seismic activity on one side of the Pacific have caused devasta-
tion on the other side, as in the case of the 1960 earthquake in Chile which
caused serious damage there and also in Japan nearly 20,000 km away.

Tsunamis behave in the open sea just like other surface waves but because of
their long wavelength, of the order of 200 km, they behave as shallow-water
waves even in the deep ocean because the ratio $h/L \sim 1/50$. Their calculated
amplitude on the deep ocean is of the order of 1 m and so they are of no
significance to ships there. It is only when they slow down and peak up near
shore that they become dangerous. The effect observed is an abnormal rise
and fall of sea level of up to several metres amplitude and period of about
15 min. Sometimes there is an initial rise, sometimes an initial fall, and
the unusual oscillations may continue for hours, occasionally for a day or
two. During the abnormal fall of sea level, ships may be left aground, tip
on their side and be swamped on the succeeding rise. During an abnormal rise
there will be little significant effect at a steep cliff shoreline, but
where the shore is flat and only slightly above normal high tide level, the
sea may pour across the flat lands and sweep away buildings or carry ships
inland and strand them there. Refraction effects also play a significant
role near shore and make certain ports particularly susceptible to damage,
while others may be less so.

Tsunami amplitudes are measurable along a coast from tide gauge records, by
special tsunami recording gauges, and from surveys of damage caused. There
have not yet been any reports of direct measurements of the amplitude of
tsunamis in deep water but gauges for this purpose are in experimental use.

In the Pacific Ocean, where most tsunamis occur because of the large number of
earthquake zones, there is a Tsunami Warning System in operation based in
Hawaii and with input from many countries around the Pacific. For this
System, seismological observatories around the ocean provide information to a
Centre near Honolulu within about one half-hour of any earthquake occurring
under or near the sea. From the earthquake epicentre information, if a
tsunami is generated its time of arrival at any point can be calculated to an
accuracy of a few minutes because the depth distribution in the Pacific is
sufficiently well known for this purpose. Because it is not yet possible to
predict whether or not a tsunami will be generated, the procedure is to alert
observing stations along the coast on either side of the epicentre and they
report to Hawaii immediately if a significant wave is observed. In this case,
all other countries in the System are warned of the possibility of a tsunami
arriving and they can take whatever precautions have been planned. If no
significant waves are observed, the alert can be cancelled.

Fig. 12.11.   Internal waves at the boundary between two layers of water of
              different densities.  Note that $h_1$ and the wave height are
              exaggerated for clarity.


INTERNAL WAVES

Thus far we have considered so-called 'surface' waves occurring at the air/
water interface.  Similar waves can occur at surfaces between different
density layers within the sea, i.e., internal waves, because the density
difference leads to a gravitational or hydrostatic pressure (caused by gravity)
restoring force if fluid is displaced vertically.  Particular surfaces are in
the thermocline in oceanic waters, where the density difference is chiefly
due to temperature difference, or at the halocline in coastal waters where
the density difference is mainly due to a salinity difference.  Of course,
the water movements are not limited to the interface itself, but extend
through the water above and below it.  For surface waves, the density of air
is so small compared with that of water (ratio about 1/800) that the former
could be ignored and the air density did not appear in the formulae for wave
speed.  However, for internal waves, the densities of the two water layers
are nearly the same.  Theory indicates, for example, that for a relatively
thin layer, $h_1 ( < L_2/20)$, of water of lower density $\rho_1$ over a deep layer,
$h_2 ( > L_2/2)$ of water of density $\rho_2$ (Fig. 12.11), the speed of an internal wave
of length $L_2$ is:

$$C_2 \simeq \left[ g \cdot h_1 \left( \frac{\rho_2 - \rho_1}{\rho_2} \right) \right]^{1/2} . \qquad\qquad (12.4)$$

For example, for coastal waters with $S_1 = 0‰$ (fresh water), $S_2 = 30‰$,
$T_1 = T_2 = 10°C$ and $h_1 = 5m$, then $C_2 = 1.1\ ms^{-1}$, while for the open sea with
$S_1 = S_2 = 35‰$, $T_1 = 25°C$, $T_2 = 20°C$ and $h_1 = 50\ m$, then $C_2 = 0.8\ m\,s^{-1}$.  These
speeds are much less than those for surface waves, a typical feature of internal waves.  At the same time, the small density difference between layers
(23.4 kg m$^{-3}$ for the coastal case and 1.4 kg m$^{-3}$ for the open sea case above)

permits the height of the internal waves to be quite large even for a small energy content. (In the open sea, the condition $L_2 > 20h_1$ may not be satisfied for all the waves and equation 12.4 must be modified for the shorter waves ($L_2 < 20h_1$); these waves are also dispersive.)

The theoretical analysis leading to equation 12.4 indicates that there will also be a wave system at the sea surface associated with the internal waves (and independent of any wind waves at the surface) but the amplitude of this surface system is normally too small to see and difficult to measure. However, the presence of the internal waves can often be detected visually by secondary effects if the upper layer is not very thick. As the internal waves travel along, the upper layer alternately gets thicker and thinner. Thus there are convergences and divergences in this layer. The convergences may cause bands of irregular ripples on the sea surface above them (perhaps by compressing short surface waves making them steeper and more visible), while there is smoother water over the divergences. The ripples are positioned just behind the crests of the internal waves. In other circumstances, generally when the upper layer is somewhat thicker, the convergence in the upper layer brings together organic material on the surface of the sea and thus changes its surface tension, tending to suppress any ripples which would be formed by a light wind at the surface. The result is a smooth band over the convergent area with wind ripples elsewhere. Thus, if a pattern of alternate bands of smooth and rippled water is seen on the sea surface, it is quite probable that there is an internal wave train below the surface. Another characteristic which may reveal the presence of internal waves is light-coloured silt in the upper layer, as from a river. If the upper layer is thin (a few metres) then over the troughs of the internal waves there will be a thicker layer of silty water which looks lighter in colour than the thinner layer over the crests of the internal waves (which generally have clear, dark water below).

In the upper layers of coastal waters, internal wave periods of 1 to 3 min. and amplitudes of several metres are often observed; in ocean waters where the density differences are smaller, periods of up to 12 hours and amplitudes of 10 to 30 m or more have been recorded.

The above discussion has been of the simple two-layer situation in which there is assumed to be a relatively thin layer of water of uniform lower density over a deep layer of uniform higher density, leading to a simple internal wave system. This model is quite realistic for coastal regions where there is a significant river runoff giving rise to an upper layer of low salinity over a deep layer of much higher salinity, with a steep gradient of salinity (i.e., density) between them. However, in most of the oceans, the density varies less abruptly with depth. In such cases internal waves still occur but the analysis is more complicated and a broader spectrum of internal waves with a greater range of periods and amplitudes may be present. (A range of periods may be present in the coastal case too, but the surface manifestations often show a fairly simple wave pattern.) Also, with a more complicated vertical density distribution the variation of the internal wave amplitude with depth may become much more complex than the variation which occurs in the simpler two-layer situation. As noted in Chapter 5, the Brunt-Väisälä frequency, N (equation 5.11), gives an upper limit for the internal wave frequency in the case of continuous density stratification (or a minimum period of $2\pi/N$).

Knowledge of the causes of these internal waves is incomplete.  Possible
causes are instability of flow where there is a strong vertical shear of
velocity (e.g., strong tidal currents through passes or over bottom irregularities), and inverted barometer effects associated with moving low atmospheric pressure systems, and related short-period variations in wind stress.

Because of the associated vertical oscillations of the water, internal waves
can be a considerable nuisance when one is attempting to determine the steady
state distribution of water properties.  They are also probably a significant
factor in promoting mixing between the upper and lower layers if they break to
form internal 'surf'.

### EFFECTS OF ROTATION

Consider the possible oscillatory (wavelike) motions of a thin layer of fluid
under the action of gravitational and Coriolis forces when the fluid is taken
to cover the whole earth.  This possibility was first considered in connection with tidal theory.  The possible motions fall into two classes.  In the
first class are gravity waves, which may be modified by rotation if they are
sufficiently long.  In the second class are motions, generally with periods
greater than a day, associated with the variation of the Coriolis parameter
($f = 2\Omega \cdot \sin \phi$) with latitude.  If the rotation rate goes to zero, the first
class of waves become ordinary gravity waves while the second class of waves
become steady currents.  If a lateral boundary is added, an additional special
type of wave may exist, called a *Kelvin wave* after Lord Kelvin who first
found the solution describing it.

The second class of waves are also called *planetary waves* or *Rossby waves*
since Rossby was the first to investigate them on the β-plane (where
$f = f_o + \beta \cdot y$ is used with $f_o$ and β both taken to be constant).  In the following, we shall discuss some of the basic characteristics of such waves.

### Modified Gravity Waves

For gravity waves with periods (T) approaching one-half pendulum day ($2\pi/f$),
the Coriolis terms ($f \cdot u$, $f \cdot v$) become comparable in size to the local time-
rate of change ($\partial u/\partial t$ or $\partial v/\partial t$).  For example, if u, v vary as $\sin (2\pi \cdot t/T)$,
then the local time derivatives have amplitudes proportional to $(2\pi/T) \cdot u$ or
$(2\pi/T) \cdot v$ which are equal to the Coriolis term for $T = 2\pi/f$.  For much shorter
periods (e.g., wind waves and swell and relatively short-period internal
waves) the Coriolis terms may be ignored since $T \ll 2\pi/f$.  For a horizontally
infinite ocean with f = constant, it can be shown that the free modified
gravity waves have periods $T < 2\pi/f$ for both surface and internal types of
gravity wave.  Such waves may be termed *gravitational-gyroscopic waves* to
indicate that both gravity and rotation are important.  (Sometimes the term
*inertial waves* is used when the periods are near the inertial period but to
avoid confusion with inertial oscillations (Chapter 8) the first term is to
be preferred.)  Forced gravity waves, e.g., those produced by the tidal
forcing with $T > 2\pi/f$ may occur.  Free waves are the possible motions which
may occur after some disturbance has occurred, e.g., a change of wind stress.

Let us take the ocean to be of uniform density (a valid model for a barotropic case as discussed in Chapter 7 at the beginning of the section on
dynamic stability).  The horizontal pressure gradients are then given by the
surface elevation gradients (as near the end of Chapter 9).  The equations

for horizontal motion, neglecting the advective accelerations which we take to be small, are:

$$\frac{\partial u}{\partial t} - f \cdot v = - g \cdot \frac{\partial \eta}{\partial x}$$

$$\frac{\partial v}{\partial t} + f \cdot u = - g \cdot \frac{\partial \eta}{\partial y} \qquad\qquad (12.5)$$

while the equation of continuity with the time average depth, h, constant and $\eta \ll h$ is:

$$\frac{\partial \eta}{\partial t} = - h \cdot \left(\frac{\partial u}{\partial x} + \frac{\partial v}{\partial y}\right) \; . \qquad\qquad (12.6)$$

Taking f = constant, a solution for a wave (which will be a long wave since T is large) travelling in the x direction is:

$$u = [(g/h)/(1 - s^2)]^{1/2} \cdot \eta_o \cdot \cos 2\pi(t/T - x/L)$$

$$v = [(g/h) \cdot s^2/(1 - s^2)]^{1/2} \cdot \eta_o \cdot \sin 2\pi(t/T - x/L)$$

$$\eta = \eta_o \cdot \cos 2\pi \cdot (t/T - x/L)$$

where $s = f \cdot T/2\pi$. The phase speed is:

$$C = [g \cdot h/(1 - s^2)]^{1/2} \; .$$

Clearly s < 1 or T < $2\pi/f$ (one-half pendulum day) for physically possible solutions. We see that the phase speed is increased by rotation and a horizontal component of motion perpendicular to the direction of propagation is introduced. This latter behaviour should not be too surprising. As a particle moves forward it will tend to be deflected to the right (or left) by the Coriolis force and vice versa as it moves back. When the period is comparable to $2\pi/f$, the Coriolis effect will be important; when the period is much shorter it will not. Likewise, internal waves with periods comparable to $2\pi/f$ will be affected by the rotation.

## Kelvin Waves

If a lateral (vertical) boundary exists, then the Kelvin wave solution to equations 12.5 and 12.6 is possible. It is, with the boundary parallel to the x axis (i.e., east-west):

$$u = (g/h)^{1/2} \cdot \eta \; , \quad v = 0, \quad \eta = \eta_o \cdot \exp(-f \cdot y/C) \cdot \cos 2\pi(t/T - x/L)$$

with $C = \sqrt{g \cdot h}$ and h = constant. In general, Kelvin waves propagate forward with the boundary on the right in the northern hemisphere (and on the left in the southern hemisphere). The amplitude is greatest at the boundary and decays exponentially away from it (a characteristic which identifies it as a type of *boundary wave*). At each point at any time, the Coriolis force balances the pressure force due to the surface slope. Kelvin waves may also occur along the equator, where f changes sign, propagating from west to east, and over abrupt changes in bottom topography (where they are called 'double' Kelvin waves because there is motion on both sides of the depth change).

## Planetary or Rossby Waves

These are waves of long period which are associated with the 'β-effect', the variation of the Coriolis parameter f with latitude. Here we shall indicate some properties of the solutions on a β-plane, leaving more complete discussions for more advanced texts (e.g., LeBlond and Mysak, 1978).

First we should consider how f variations with latitude may lead to oscillatory motions. When discussing equatorial under-currents in Chapter 9 we pointed out how an eastward current might oscillate if perturbed. More generally, suppose that we move northward a parcel of water whose initial relative vorticity is zero, with no effects such as friction or depth changes which will cause its potential vorticity to change. As f increases, the parcel will have negative relative vorticity when displaced (northward) and will circulate clockwise. Because of its variation, the Coriolis force will be a maximum in size on the poleward part of the parcel and a minimum on the equatorward part, the variation of f thus leads to a net southward force tending to produce southward displacement, i.e., a restoring force; if this force pushes the parcel south of the latitude of zero relative vorticity (overshoots) the circulation becomes anticlockwise, and considering f variations the parcel now has a net northward force, i.e., again a restoring force. Thus the variation of f provides a restoring force (in the horizontal) allowing oscillations to occur just as the effect of gravity does (in the vertical) for surface or internal waves. The flow will be nearly horizontal and for $T \gg f/2\pi$ will be essentially geostrophic (often termed 'quasi-geostrophic') since the time derivatives (and advective accelerations) will be small compared with the Coriolis terms. Thus with sufficiently detailed observations, such flows (or waves) may be shown by geostrophic calculations. Assuming the flow to be non-divergent in the horizontal and barotropic (depth independent) it can be shown that:

$$C_x = - \beta \cdot L^2/4\pi^2 \qquad\qquad\qquad (12.7)$$

where $C_x$ is the phase velocity in the x direction (east-west) and L is the wave-length. The minus sign indicates that the phase velocity is always in the minus x direction (to the west) although the group velocity (direction of energy propagation) may be in any direction. If the wave is moving westward with crests north-south then the period $T = L/C_x = 4\pi^2/(\beta \cdot L)$. The period increases as L decreases in contrast to surface gravity waves. This result remains true for wave crests in any given direction; the shorter waves are of longer period. For $\beta = 2 \times 10^{-11} \, m^{-1} s^{-1}$ and L = 1,000 km, $C_x \approx -0.51 \, m \, s^{-1}$ and with crests north-south, $T \simeq 23$ days.

If we take the divergence into account (or equivalently do not neglect the surface elevation) equation 12.7 becomes

$$C_x = - \beta/[4\pi^2/L^2 + f^2/(g \cdot h)] \quad ;$$

thus the divergence effects are only important for the longer wavelength (shorter period) waves. With $f \simeq 10^{-4}$ (about 45° latitude), h = 4 km and L < 2,000 km, then $f^2/(g \cdot h) \simeq 2.5 \times 10^{-13}$ and may be ignored relative to $4\pi^2/L^2 > 100 \times 10^{-13}$.

The mesoscale 'eddies' observed in the POLYGON and MODE experiments may be interpreted as linear superpositions of Rossby waves (including internal types). If such interpretations are correct then the eddies are nearly linear phenomena and would have little effect on the large-scale mean flow. Eddies may be generated by baroclinic or barotropic instability. If non-linear effects are important they tend to grow in size (behaving like two-dimensional turbulence). However, if they grow sufficiently they may become nearly linear (a superposition of Rossby waves) *before* they interact with the mean flow and will radiate their energy away (perhaps to be dissipated or reflected at boundaries). If they grow to the size where the dynamics are linear before interacting with the mean flow they may not be very important for the mean flow except perhaps as an energy sink. Since the eddies seem to start out rather small in the ocean, compared with those in the atmosphere, the eddies may not be very important for the mean flow except as a loss process in contrast to the atmospheric case. Their importance has not yet been established, as noted before in Chapter II.

Topographic Effects

More generally, it is not f variations which matter but variations of f/D. When depth variations dominate (as often occurs near coasts) the possible Rossby waves are termed *topographic*. Indeed, in laboratory scale models, where the rotation rate cannot be varied with position, the β-effect can be simulated by varying D.

Variations in topography may lead to wave trapping (concentration of wave energy in certain regions). (Variation of f may also lead to trapping, particularly near the equator.) Trapping may occur with gravity waves for which the term *edge waves* is used or with Rossby-type waves for which the term is *shelf waves*. Kelvin waves may also be considered a special case; their very existence depends on the presence of a boundary (or on f changing sign at the equator or on an abrupt depth change). They are a boundary phenomenon in the sense that the oscillations are large near the boundary and decay away from it.

As an example, consider gravity waves approaching a shore at an angle (as in Fig. 12.5) with the depth decreasing but with a vertical boundary (cliff) so that the depth does not go to zero. By refraction, the wave crests will become more nearly parallel to the shore. If they do not break they will be essentially totally reflected at the cliff at an angle equal to the incident angle.

They will then travel outward, with refraction causing the crests to become perpendicular to the shore and then will turn inward again to be reflected once again. The wave orthogonals will form a series of arcs as in Fig. 12.12. For planetary waves, f/D variations may produce similar behaviour.

OCEAN

CLIFF

Fig. 12.12.    Refraction leading to wave trapping when reflection occurs
before wave breaking.   One wave orthogonal is shown.

# CHAPTER 13
# **Tides**

## INTRODUCTION

The *tide* is the name given to the alternating rise and fall of sea level with a period of about 12½ hours (about 25 hours in a few places). The rise and fall is the most obvious feature to most observers but fundamentally the prime phenomena are horizontal tidal motions (currents); the rise and fall at the coast is simply a consequence of the convergence or divergence occurring there when the tidal currents flow toward or away from the shore. This aspect of tidal theory, vertical versus horizontal motion, will be discussed in more detail later.

It should be noted that tidal movements occur in the atmosphere and in the 'solid' earth as well as in the sea but we will only be concerned with the oceanic tides in this text.

## TIDE-PRODUCING FORCES

Tides are a consequence of the simultaneous action of the moon's gravitational force, the sun's gravitational force, and the revolution about one another of the earth and moon and the earth and sun. In principle, the other planets in the solar system also exert tidal forces on the earth but their values are so small compared to those of the moon and sun that they are quite negligible.

The magnitude of the total tide-producing force is only of the order of $10^{-7}$ times that due to earth's gravity but as it is a body force, acting on the total mass of the ocean, and has horizontal components, it is a very significant one.

We will follow the procedure suggested by Darwin (1911, see the Further Reading list) to explain the salient characteristics of the tide-producing forces, looking at the earth/moon pair first.

If we imagine ourselves away from the earth and looking down on it and its moon from above the north pole, the relative arrangement will be as in Fig. 13.1a which is not to scale, the moon having been brought close to the earth for convenience. From this point of view the earth will appear to rotate anti-clockwise about its axis and the earth/moon pair will also rotate anti-clockwise like an assymetric dumbbell. As the centre of mass of the earth/moon pair is about one-quarter earth radius inside the earth, the rotation of the pair will be about an axis at the position Z and perpendicular to the paper. To understand the effect of the moon's gravitation plus the earth/moon rotation, we will assume for the moment that the earth is not

Fig. 13.1.  (a) Earth-moon pair rotates about a common centre of mass at Z,
(b) successive positions of earth-moon pair with earth's diurnal
rotation suppressed, (c) motions of two points on earth's surface.

rotating on its axis.  Fig. 13.1b shows the character of the motion now in
four stages (i) to (iv) of one complete rotation of the moon about the axis
through Z.  The motion of the earth now is an unusual one in that its *orientation* remains fixed (the face is drawn to emphasize this characteristic)
but every part of the earth rotates in a circle of radius equal to the distance from the earth centre to the axis Z.  (The easiest way to observe such
a motion is to place your hand flat on a table, fingers outstretched, and to
move it so that one point, e.g., the tip of the thumb, rotates in a horizontal circle of about 10 cm diameter, keeping your forearm pointing in the same
direction all the time.  You will see that every part of your hand describes
the same sized circle, fingertips and palm alike).  Fig. 13.1c shows the
circles described by points A and B on the earth (Fig. 13.1b).  For all points

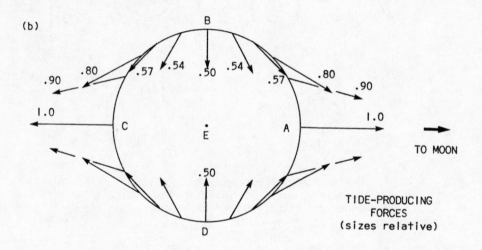

Fig. 13.2.   (a) Directions of centripetal force per unit mass (CF) and moon's
gravitational force per unit mass (F) at points on earth (not
to scale), (b) directions of residuals of CF and F at various
points on the earth's surface and their magnitudes relative to
values at A or C taken as unity.
(Correct to first order – see text.)

on the earth to move in a circle of this size it is necessary for there to be
a centripetal acceleration (or force/unit mass) of the same magnitude everywhere and directed parallel to the earth/moon axis toward the moon. This
centripetal acceleration is, on average, provided by the gravitational
attraction of the moon but as this attraction is not quite uniform over the
surface of the earth, there is a slight excess of gravitational acceleration
on the near side to the moon and a slight defect on the opposite side. (The
gravitational acceleration is not uniform because the distance from the moon
to points on the earth's surface varies from place to place).

Fig. 13.2a shows the earth/moon pair, with the moon close to the earth to
emphasize some of the angles for convenience in showing components of forces
(per unit mass, i.e., accelerations) later. (In the following, the term
'force' will be taken to be per unit mass without stating so explicitly every
time.) We will consider the forces at four points on the earth's surface, A
and C which are the nearest and farthest points from the moon, and B and D
which are equidistant from the moon. At points B and D, the gravitational
force of the moon on unit mass on the earth is shown by arrows $F_b$ and $F_d$, respectively. These two forces will be equal in magnitude because of their
equal distance from the moon. Because they occur at very nearly the same
distance from the moon as is the centre of the earth (remembering that the
angle BME is grossly exaggerated in Fig. 13.2a), the value of this gravitational force will be almost exactly equal to the average value for the force
of the moon on a unit mass on earth ($F/M_e$), given from Newton's Law of Gravitation by: $F/M_e = G \cdot M_m/r^2$, where G is the Gravitational Constant (= 6.67
$\times 10^{-11} N kg^{-2} m^2$), $M_e$ and $M_m$ are the masses of the earth and moon respectively,
and r is the distance between their centres. The moon's gravitational force
per unit mass ($F_a$) at point A will be larger than the average value $F/M_e$
because it is nearer to the moon, while the value at C($F_c$) will be less than
$F/M_e$. The differences are about ± 3%. The values of the four forces $F_a$, $F_b$,
$F_c$, and $F_d$ are shown semi-quantitatively by the full-line arrows in Fig.
13.2a. The total centripetal force which keeps the earth and moon at their
constant distance apart while rotating is provided by their gravitational
attraction and so must be equal to the value F, or $F/M_e$ per unit mass (= CF),
as is shown in Fig. 13.2a by the dashed arrows.

At D, by Newton's law of gravitation, the force/unit mass $F_d = G \cdot M_m/(r^2 + R^2)$
where R is the radius of the earth. Rewriting the denominator as $r^2(1 + R^2/r^2)$
and expanding in series gives $F_d = (G \cdot M_m/r^2) \cdot (1 - R^2/r^2 + \cdots)$. As $R/r \simeq 1/60$,
$F_d$ is very nearly equal to CF in size, the difference being 1 part in 3,600.
Resolving $F_d$ into components parallel and perpendicular to the line EM, i.e.,
$F_{dx}( = F_d \cdot r/(r^2 + R^2)^{1/2})$ and $F_{dy}( = F_d \cdot R/(r^2 + R^2)^{1/2})$ respectively, we get,
again using series expansion:

$$F_{dx} = (G \cdot M_m/r^2) \cdot [1 - (3/2) \cdot (R^2/r^2) + \cdots] \text{ and}$$

$$F_{dy} = (G \cdot M_m \cdot R/r^3) \cdot [1 - (3/2) \cdot (R^2/r^2) + \cdots] .$$

To an accuracy of 1/2400, we may take $F_{dx} = CF$ and the residual to be $F_{dy} \simeq$
CF/60 directed *inward*. (Note however that the size of ($F_{dx} - CF$) is about 1/40
of $F_{dy}$.) Similarly at B there will be the residual force $F_{by}$ of exactly the
same size as $F_{dy}$ also directed *inward*. At A, $F_a = G \cdot M_m/(r - R)^2$ is directed
*outward* (toward the moon). Expanding we have:

$$F_a = (G \cdot M_m/r^2) \cdot (1 + 2R/r + 3R^2/r^2 + \cdots), \text{ and the residual}$$

$$F_{ax} = F_a - CF = (2G \cdot M_m \cdot R/r^3) \cdot [1 + (3/2) \cdot (R/r) + \cdots] \text{ directed } \textit{outward}.$$

To first order (i.e., neglecting terms of $O(R/r)$ compared to unity) $F_{ax} = CF/30$ = $2F_{by}$. The neglected term is about 1/40 of $F_{ax}$. At C the moon's gravitational force $F_C$ will be slightly less than needed to provide CF so that a mass there will behave as though there were an *outward* force

$$F_{cx} = F_c - CF = G \cdot M_m/(r+R)^2 - CF = -(2 \cdot G \cdot M_m \cdot R/r^3) \cdot [1 - (3/2) \cdot (R/r) + \cdots].$$

Again to first order, $F_{cx} = -CF/30$ although its size is really about 2.5% smaller than CF/30 or about 5% smaller than $F_{ax}$. For our purposes, to illustrate the principles, we need only use the residual forces correct to first order. (The higher order terms can be retained for more detailed calculations if necessary. Remember also that although the forces are really slightly assymmetric, $F_{ax}$ being slightly larger than $-F_{cx}$, the sum of the residuals (integral) over the whole earth's surface is zero. This result is obvious for the direction perpendicular to EM because for each point there is a corresponding point where the force component on a unit mass perpendicular to EM is exactly equal and opposite, e.g., points B and D. In the parallel direction, the result is not so obvious and it is left as an exercise for the reader with the necessary mathematical background to show that the integral of the force components in this direction proportional to $R^2/r^4$ does vanish.)

Between points B or D and A or C, the residual force will gradually change from an inward direction to an outward direction. Fig. 13.2b shows the directions of the residual forces for a series of points on the earth's surface in the plane of the drawing, and shows their magnitudes relative to the values at A or C taken as unity. The distribution of the residual forces over the rest of the surface of the earth can be obtained by rotating Fig. 13.2b about the EM axis. These residual forces, then, are the *tide-producing forces*.

What we have as the result of the analysis so far is a pattern of residual forces directed outward from the earth's surface over areas on the near *and* far sides to the moon, but directed inward in between these areas. This pattern is tied to the earth/moon axis, but the earth has been assumed not to be rotating. If we now restore the rotation of the earth about its polar axis, each point on the earth's surface will, in one day, pass through the whole pattern of residual forces and so experience a cycle of tide-producing forces as in Fig. 13.2b with *two* passages of outward forces and *two* passages of inward forces. That is, the tide-producing forces will have a period of one-half day even though the earth has a rotation period of one day. This is the basic reason for the existence of semi-diurnal tides for a diurnal rotation of the earth. (The day here is the lunar day of about 24.8 h because the moon is also moving with a 27.3 day period around the earth compared with the earth's rotation in 24.0 h relative to the sun.)

## Components of the Tide-Producing Forces

The earth/sun pair sets up a similar pattern of forces to that for the earth/moon pair but it differs from the latter in two respects: the maximum effect due to the sun is only about one-half of that due to the moon (because its greater distance outweighs its greater mass), and as the sun and moon do not rotate in synchronism the force patterns rotate independently and hence give rise to a rather complicated resultant.

The facts that the paths of rotation of the sun and moon about the earth are
not circles but ellipses, and that the planes of rotation are not in the
equatorial plane but move north and south with an annual cycle for the sun
and a monthly cycle for the moon add further complications to the resultant
tide-producing forces.  However, the motions of the sun and moon are known
very exactly and it is possible to express the resultant tide-producing forces
as the sum of a number of constant *simple-harmonic components* (sine waves),
each of which has its own characteristic (constant) amplitude, phase and
period of fluctuation.  Some of the more important components with their size
relative to the largest, the principal lunar force, taken as 100 are given in
Table 13.1:

TABLE 13.1    Characteristics of Some of the Principal Tide-
Producing Force Components

| Name | Symbol | Period (Solar Hours) | Relative Size |
|------|--------|----------------------|---------------|
| Semi-diurnal: | | | |
| Principal lunar | $M_2$ | 12.4 | 100 |
| Principal solar | $S_2$ | 12.00 | 47 |
| Larger lunar elliptic | $N_2$ | 12.7 | 19 |
| Luni-solar semi-diurnal | $K_2$ | 11.97 | 13 |
| Diurnal: | | | |
| Luni-solar diurnal | $K_1$ | 23.9 | 58 |
| Principal lunar diurnal | $O_1$ | 25.8 | 42 |
| Principal solar diurnal | $P_1$ | 24.1 | 19 |
| Larger lunar elliptic | $Q_1$ | 26.9 | 8 |
| Long period: | | | |
| Lunar fortnightly | $M_f$ | 328 | 17 |
| Lunar monthly | $M_m$ | 661 | 9 |
| Solar semi-annual | $S_{sa}$ | 2191 | 8 |

There are up to 65 components which are recognized as significant in some
circumstances, e.g., in describing tides in river estuaries.


OCEAN RESPONSES TO THE TIDE-PRODUCING FORCES - TIDAL THEORIES

There are two tidal theories, the equilibrium theory and the dynamic theory.

In the *equilibrium theory*, the *vertical* components of the tide producing
forces are regarded as the significant ones and the ocean surface is supposed
to be lifted where the resultant force is outward from the centre of the
earth, i.e., upward, and to be depressed where the resultant force is downward (*cf.* Fig. 13.2b).  Then, as the force patterns revolve with the moon and
sun, so do the high and low waters, to produce the observed tides.  However,
the rise and fall predicted by this theory are too small (about 0.35 and
0.18 m respectively) compared to the observed tides.  Also the theory, in
effect, requires the mass of the water to be subject to gravitational attraction but to have no inertia, that is, to respond instantaneously to the tide-
producing forces - a very unreal situation.  This theory may be satisfactory

for application to tides in the solid earth but for the oceans it has been replaced by the dynamical approach.

The *dynamic theory* argues that the *horizontal* components of the tide-producing forces are more important, since water is a fluid and can be moved in the horizontal direction with relatively little effort. The water flows in response to the tide-producing forces, forming tidal currents in the whole of the ocean. In effect, the tide-producing forces generate 'forced' waves in the ocean. (These are motions which follow the period of the acting force rather than having a periodic motion determined by the mass and restoring force of the system itself.) The rise and fall at the boundary is simply a result of the flow convergence or divergence there. In principle, then, one could apply the equations of motion and calculate the tidal currents anywhere in the ocean, and the rise and fall at the shore. Unfortunately, the real ocean basins are very complicated in shape and it is not possible to obtain exact solutions to the equations of motion. However, it is possible to solve the equations numerically and promising results have been obtained for simplified models of the ocean basins. The limitations to obtaining detailed solutions lie in the large amount of high-capacity computer time required. Whether refinements of this direct approach will be practical, or whether novel approaches may prove less laborious, is not known at the present time.


## THE PRACTICAL APPROACH TO TIDE PREDICTION

If present theory is inadequate to predict tides from dynamic principles, how is it that one can buy, for a modest sum, books of *tide tables* (predictions of times and heights of the tides and in some places for tidal currents) for a large proportion of the ports of the world? In effect, one uses the ocean itself empirically as a computer to solve the equations of motion. One records the rise and fall of the water as a function of time at a particular location for a period of time, analyses the resulting tide height curve for its component characteristics and uses them to perform the prediction into the future. This procedure was used empirically even before the development of the present tidal theory.

The recorded tide curve is a complex harmonic, i.e., the sum of many simple harmonics of different periods and amplitudes. It is resolved into its simple harmonic 'constituents' by mathematical procedures which are straightforward (but tedious if done by hand). Each constituent can be represented by a sine curve with its own period and phase, whose amplitude represents its contribution (above or below mean sea level) to the total tide. These constituents can be drawn for as far into the future as we wish, and then for any future time we can determine the expected tidal height simply by adding together all the constituents for that time. In practice, our observations of the tide curve have a limited accuracy so there is a limit to the accuracy with which we can determine the constituents and our predictions into the future become less accurate the further ahead in time that we go. Naturally, the longer the series of observations that we have, the more accurate are likely to be the constituents and therefore our predictions.

Although tidal theory itself cannot yet predict the tides to a satisfactory accuracy, it can tell the practical man what constituents to look for, each constituent corresponding to one of the components of the tide-producing forces (and perhaps components at sum and difference frequencies which arise when non-linear effects become important, usually in coastal areas). It was this contribution from tidal theory which changed tidal prediction from a purely empirical procedure to one based on sound physical principles.

Until the recent past, the *harmonic analysis* of the recorded tide curve into its constituents was carried out by paper-and-pencil methods on tabulations of hourly water heights, but it is now done by digital computer. The prediction into the future was carried out by an analogue device, invented by Lord Kelvin, in which the constituents were represented by rotating eccentrics (cams) whose throw was proportional to the amplitude and whose rate of rotation was inversely proportional to the period. A steel tape passing over all the cams totalled their displacements and this total was recorded on a paper strip, the result being a tidal height versus time curve for the future. This step is now carried out by digital computer, which is programmed to compute the times and heights of high and low waters and to assemble them into tabular form with appropriate headings, etc., ready for printing.

An important point to note here is that the tide *height constituents* observed at a particular locality do not necessarily have the same relative proportions as the tide-producing *force components*. The particular shape of the ocean basin in the vicinity causes the water to respond more readily to some components than to others, which is the main reason for the differences between tides in different parts of the ocean. The purpose of recording the tide for a period is effectively an analogue procedure for determining the local response, when our theoretical techniques are not strong enough to do so.

The number of constituents used for prediction depends on the accuracy required. Often, the use of the first seven in Table 13.1 (i.e., $M_2$, $S_2$, $N_2$, $K_2$, $K_1$, $O_1$ and $P_1$) will be sufficient to predict the tide within about 10%, but generally 20 to 30 constituents are used for predicting two or three years ahead for ports close to the ocean and sixty or more for those in river estuaries where the tide is rendered more complicated by bottom topography.

It is usual to use a continuous tide record for 369 days for analysis into 20 to 30 constituents, and to repeat the analysis for several 'years' to improve accuracy. Analysis for fewer constituents may be done with as few as 29 or even 15 days' continuous records. The number of constituents which can be calculated (or 'resolved' as the process is often termed) depends on the record length. The reader needing more detailed information on the practical limits of tidal analysis should consult a text on the subject (e.g., that by Godin in the Further Reading list).

It should be remembered that sea level is affected by other factors as well as the tide-producing forces, e.g., atmospheric pressure and wind set-up. These so-called meteorological tides are left as a residual by the harmonic analysis process and cannot be predicted (unless they have a tidal period, e.g., land-sea breeze effects which will be included as part of the $S_2$ tide). In the absence of such disturbances, the tide predictions are generally accurate to about ±3 cm and ±5 min. Meteorological effects, leaving out extremes such as hurricanes, may cause differences of tens of centimetres and tens of minutes.

## THE MEASUREMENT OF TIDES

The simplest procedure for recording tides is to mount a vertical scale on a pier or wharf and to note visually the height of the water at, say, hourly intervals for long enough to obtain a record suitable for analysis. This is a tedious procedure.

The great proportion of tide records have been obtained with a float-type recorder. A stilling well is mounted in the water with a recorder on top of it. The stilling well is a vertical pipe with only a small hole at the bottom, below the lowest water-level, so that the effects of waves of periods much shorter than tidal are damped out and the rise and fall of water in the well follows chiefly the tidal rise and fall. (Longer period waves, such as tsunamis, will be recorded to some degree.) A float on the water in the well is connected by a wire to a pulley which drives, through a reducing mechanism, a pencil which then moves back and forth parallel to the axis of a drum, carrying paper, which is rotated by a clockwork drive. The pencil then records on the paper a graph of tide height versus time.

In other instruments, a pressure sensor is mounted in the water below low tide level and connected to a shore-mounted instrument which records the variations of hydrostatic pressure with time. The pressure can then be converted to water depth. Suitable damping is used to make the sensor or the recorder insensitive to waves. In the bubbler-type gauge, a tank of air under pressure is connected through a pressure-reducing mechanism to a pipe whose open end is fixed in the water below low tide. The shore instrument then measures the air pressure needed to just cause air to bubble out of the open end of the pipe, thus measuring the water pressure there and hence the water depth. Again, a record of pressure versus time yields the desired tide height curve. The advantage of this type of tide gauge over the remote pressure sensor type is that the instrumentation is on shore and accessible for servicing. All that is in the water is a length of hose.

For use in cold regions, where sea ice in winter might damage any structure mounted through the water surface, the pressure sensor type may be used with shore recording, the connecting cable being buried in a trench. For severe ice conditions, a self-contained pressure/time recorder can be mounted on the sea bottom, being placed in position during one ice-free season and re-covered during the next. These self-recording instruments are now being used also to record tides over the continental shelf and slope, on seamounts and even in the deep sea.

The tidal *ranges* (vertical differences between successive high and low waters) to be measured vary from almost zero, e.g., in places in the Faeroes, to 15 m, in the Bay of Fundy in Canada.

## TYPES OF TIDES

The simplest classification of tides uses, as the distinguishing feature, the emphasis on response to the diurnal or to the semi-diurnal components of the tide-producing forces. In Fig. 13.3 are shown the two basic types with varieties of the second. For *diurnal* tides (Fig. 13.3a) there is one high water and one low water in each lunar day (about 24.8 h), while for *semi-diurnal* tides (Fig. 13.3b,c) there are two high and two low waters in the same time

Fig. 13.3.   Simple classification of tides as:  (a) diurnal, (b) semi-diurnal
             (equal), (c) semi-diurnal (unequal).  HW - high water, LW - low
             water, HHW - higher high water, LLW - lower low water, LHW -
             lower high water, HLW - higher low water.


interval.  For semi-diurnal tides, in some regions two successive high waters
will have nearly the same height and two successive low waters will have
nearly the same (lower) height (Fig. 13.3b, *semi-diurnal equal tides*); in
other regions, successive high waters and successive low waters will each
have different heights (Fig. 13.3c, *semi-diurnal unequal tides*).  In some
locations, a predominantly semi-diurnal tide becomes diurnal for a short time
each month during neap tides.

As the forces due to the sun and moon come into phase, the range of the tide
increases to a maximum (*spring tides*).  This maximum occurs when the sun and
moon are both on the same or on opposite sides of the earth.  When the sun
and moon are nearest to 90° to each other the resultant forces have their
minimum value and the tides have their minimum range (*neap tides*).  Successive
spring or neap tides occur at intervals of about 15 days.

Often the same type of tide is found for long distances along a coast so that
a tide record at one port in the region will be sufficient to determine the
type of tide.  The main differences to be expected are in the phase and
amplitude of the tide at other points in the region.  It is therefore sufficient to collect long-term records at a few (principal) points, usually
ports, to determine the important constituents and then to make shorter term
observations at subsidiary points to determine the relative phases (times of
high or low water relative to those at the principal points) and relative
tidal ranges.  This procedure works for an open coast with simple bottom
topography.  Along a complicated coast, such as that of British Columbia or
of southern Chile, or in an island archipelago, it may be necessary to have
the principal ports for long-term observations closer together.  Only measurements in the field can determine just how close together or far apart need be
the recording stations.

For practical reasons, almost all of our information about tidal rise and
fall is for the coast, because here there are fixed structures to which one
can mount tide gauges of a simple mechanical type which have good reliability
and will record unattended for long periods of time.  In the last ten years
or so, considerable effort has been expended in obtaining tide records in
water of some hundreds of metres depth on the continental shelf or slope.
These are for periods of a few weeks.  For the deep ocean, there are at present just a few records obtained with self-contained instruments of or similar
to the type designed for recording tsunamis in deep water.

## TIDAL CURRENTS

Although tidal currents are the basic phenomenon, they have been less studied
than the tidal rise and fall.  There are two reasons.  The instrumentation required is rather more complicated and less likely to operate satisfactorily
for long periods unattended, and secondly the current characteristics may
vary markedly over distances of hundreds or even tens of metres horizontally
and over depths of only a few metres.  In consequence, and for practical
reasons, most tidal current information is for locations in narrow passages in
shipping routes where the currents are strong and have a significant effect
on the navigation of ships.  In such locations, the tidal current speeds tend
to be about 90° out of phase with the tidal rise and fall, i.e., the maximum
current speeds will be at the middle of the rise or fall, with slack water
near high or low water, although considerable variations from this pattern
occur near complicated coastlines.

The common pattern in narrow waterways is of a *flood* current in one direction
as the tide is rising and an *ebb* current in the opposite direction when it is
falling.  However, in the more open waters of the continental shelf and in
shallow seas, the characteristics of the tidal currents are that as they vary
in speed, often never decreasing to zero, their direction rotates, usually
with a semi-diurnal period dominating.

## TIDES IN BAYS - RESONANCE

In some bays, the tidal range is very large compared with the tidal range in
the ocean near to the mouth of the bay.  This phenomenon is often attributed
to *resonance* - the water in the bay having a natural period of oscillation
close to that of the astronomical tides and therefore accumulating energy from
them.  The Bay of Fundy in eastern Canada is a frequently quoted example.

Let us examine the conditions necessary for resonance to occur.  First, consider a long, narrow body of water (Fig. 13.4a) of length $L_b$ of depth h when
the water is still, and of constant width.  For simplicity we assume that the
bottom is flat and the ends are vertical.  Such a body of water can be caused
to oscillate, and the simplest mode is one in which the water at the ends
(A, E) goes up and down parallel to the end walls (anti-nodes), that at the
middle C goes back and forth with no vertical motion (node), while that in
between, as at B and D, moves both up and down and horizontally.  The lines
1, 2 and 3 show three successive positions of the water surface.  It is quite
easy to demonstrate this phenomenon in an ordinary household bath by moving
one's hand back and forth through a few centimetres near the middle (C) or
up and down at one end.  It can also be done in a swimming pool, though

Fig. 13.4.    Tidal resonance in bays of different lengths.

here it needs a concerted effort by several people moving their bodies to get
the water to oscillate satisfactorily.  In either case it will be found that
it is necessary to apply the stimulus at a fairly specific frequency or period
to generate and maintain the oscillations.  For the household bath the period
will be of the order of two to three seconds, while for the pool it would be
of the order of ten seconds.  In the bath you could also determine that the
period of oscillation depends on the depth of water, becoming less as the
depth increases.

The reason for these so-called 'standing waves' or *seiches* is that waves
travelling along the body of water are reflected at the far end and the two
sets of waves travelling in opposite directions can interfere constructively
with each other, i.e., their amplitudes will add together, if the wave speed
and the length of the water body are appropriately related.  It can be shown
that for the simplest type of oscillation (the fundamental) shown in Fig.
13.4a then the period

$$T_f = 2L_b / \sqrt{g \cdot h}.$$                                       (13.1)

Because the length $L_b$ in this case is one-half the length of the travelling
waves, this arrangement is called a 'half-wave oscillator'.  A few values for
the period $T_f$ for various values of $L_b$ and h are shown in Table 13.2:

TABLE 13.2    Values of $T_f = 2L_b/\sqrt{g \cdot h}$ in hours for
Combinations of $L_b$ (km) and h (m)

| h | $L_b$ = 100 | 500 Period | 1,000 km |
|---|---|---|---|
| 50 m | 2.5 | 12.6 | 25.1 hours |
| 100 | 1.8 | 8.9 | 17.7 |
| 200 | 1.3 | 6.3 | 12.6 |
| 500 | 0.8 | 4.0 | 7.9 |
| 1,000 | 0.6 | 2.8 | 5.6 |

Now suppose that instead of having a body of water closed at both ends (a
'closed basin') we have one which is open to a tidal sea at one end (an 'open
basin') as in Fig. 13.4b so that water can flow in during the flood tide and
out during the ebb.  It would be possible for the shorter length bay, CE, to
behave like the right hand half of that in Fig. 13.4a and oscillate with a
specific natural period.  This body of water would be called a 'quarter-wave
oscillator' as its length $L_C$ is one-quarter of the length of the travelling
wave.   Then $L_C = 0.5L_b$ so that the natural period of oscillation

$$T_c = 4L_c / \sqrt{g \cdot h}.$$
(13.2)

The situation in Fig. 13.4b is rather artificial because we have imagined
water flowing in and out at the node C without any vertical motion.  This
arrangement might be set up in the laboratory but is not likely in nature.
A more probable situation is that in Fig. 13.4c where the end of the bay is
inside the node C so that the sea goes up and down as well as flowing in and
out at the same time.  The important feature of this arrangement is that the
vertical amplitude of surface motion (tidal range) is greater at the closed
(head) end E than at the open (mouth) end of the bay near C; in other words,
amplification of the tidal range occurs.  For such amplification, the resonant
length of the bay $L_C$ is related to the depth by equation 13.2, and for a semi-
diurnal tidal period of 12.4 h some related values of $L_C$ and h are given in
Table 13.3:

TABLE 13.3    Related Values of Length $L_C$ and Depth h for the Fundamental
Period of Oscillation of an Open Bay to be 12.4 h

| h | = | 100 | 200 | 500 | 1,000 m |
|---|---|---|---|---|---|
| $L_c$ | = | 350 | 490 | 780 | 1,100 km |

Real bays, etc., of course, do not have flat bottoms of uniform depth nor flat,
vertical ends, but it is possible to calculate the resonant length for an
irregular bay to reasonable accuracy by a step-by-step calculation to allow
for the varying depth.  However, we can compare the ideal calculations of
Table 13.3 with the dimensions of fjords (as in the coasts of British Columbia/
Alaska, Norway or Chile) using their mean depth and actual length.  We find
that for these 'bays', a typical mean depth is about 500 m with a length of
100 km.  They are therefore much shorter than the critical length for large

amplification (Table 13.3). Their dimensions are more like those of Fig. 13.4d, and only a small amplification occurs (about 5%).

The Bay of Fundy in eastern Canada is much shallower than the fjords (of the order of 100 m) and its actual length (of the order of 300 km for the Bay itself) is nearer to the critical length for resonance to the semi-diurnal tide components. It has often been quoted as an example of resonant amplification because tidal ranges of some 15 m occur near the head compared with 5 m near the mouth. Some investigators have disputed this hypothesis, calculating a resonant period as low as 9.0 h which would be too far from the semi-diurnal components for significant amplification to occur. It was suggested that the large ranges at the head were simply due to the funnelling effect of the bay toward the head, i.e., to volume continuity. At the same time we should point out that one difficulty in making a step-by-step calculation of the resonant period is to decide where to take the mouth of the bay. For a long, narrow fjord which opens suddenly into the ocean, there is not much uncertainty but for a wide-mouthed bay system, such as the Bay of Fundy together with the Gulf of Maine (for which also the axis is not perpendicular to the shoreline), there may be a temptation to select a length for the bay system which will yield a resonant period which suits one's preconceived ideas! However, C. Garrett compared the ratios of the major semi-diurnal tide characteristics inside and outside the bay system, i.e., compared the response characteristics of the system with the forcing characteristics of the ocean tide. From this calculation he concluded that the resonant period for the system is about 13.3 h which is close enough to the probable chief forcing periods (12.0, 12.4 and 12.7 h) for the resonant response explanation to be acceptable. In this example, the mouth of the bay system is determined by where the depth increases rapidly on the shelf, causing a sharp impedance change, rather than by the position of a narrow opening to the sea.

## STORM SURGES

These are mentioned here to emphasize that unusual rises of sea level may occur due to other causes than exceptionally high tides (and tsunamis as discussed in Chapter 12). Storm surges are the result of the frictional stress of strong winds blowing toward the land and causing the water level to be 'set up' by as much as several metres. This effect has caused severe flooding of low-lying areas at the south end of the North Sea during strong northerly winds and in areas at the north of the Bay of Bengal during cyclones. In the latter case, the low atmospheric pressure at the centre of the cyclone can additionally raise the sea level (the 'inverted barometer effect'). A drop of pressure of 3 kPa ( $\equiv$ 30 millibars) can raise sea level by 0.3 m. This value is the response to an atmospheric pressure change for a stationary system after a long enough time for equilibrium to be reached. The actual response may be greater or lesser than this amount depending on the topography of the area and the speed at which the storm centre moves. The only connection with the astronomical tides is that if these various other causes of increased sea level occur during a period of high tide then the disastrous effects will be compounded.

In bounded bodies of water, such as lakes or small seas, after the wind stress decreases, relaxation oscillations may take place. These are standing waves or seiches and they die out as the mechanical energy of the water is dissipated.

# CHAPTER 14

# Some Presently Active and Future Work

We have described many models which attempt to demonstrate the importance of particular dynamic effects. A model is always a simplification of the real system and attempts to include only the effects which are important in producing a particular phenomenon. Such models are useful in improving our understanding even when they do not represent the details of the real ocean. For example, Stommel's rectangular ocean model, discussed in Chapter 9, is clearly not intended to represent a real ocean; however, it does demonstrate the importance of the variation of the Coriolis parameter with latitude. The more complicated and detailed numerical simulation models described in Chapter II attempt to represent the ocean more realistically.

Because of the complicated geometry and the importance of non-linear effects, analytical modelling has not and probably cannot produce a complete ocean model. While the numerical models can be more representative and show promise, none of them, as yet, has produced a quantitatively correct model of the general ocean circulation. Both types of modelling have been used to demonstrate possible mechanisms to explain certain features of the ocean circulation. It is clear that much work remains to be done before we can be certain that we have an adequate quantitative understanding of ocean dynamics. It is also quite clear that our observational data base is inadequate. It is difficult to understand a system theoretically which one cannot describe sufficiently well.

Exploration of possible parameter ranges has been limited by the speed of available computers, particularly in models attempting to simulate the actual ocean. But the new, faster machines which have become available should help to reduce this problem. By running models of sufficient resolution that all dynamic effects are permitted, and comparing them critically with observations, it may be possible to determine which effects are important. The necessary verification observations will probably have to be collected specially for the comparisons, and obtaining them will be a non-trivial problem, to say the least. (Models of smaller regions such as rivers, estuaries and coastal seas provide examples of how such models may be adjusted to simulate a geophysical flow adequately, that is, to predict the information which we want, such as tidal currents, height of storm surges or the effects of engineering structures on the circulation. We are still a long way from having the necessary data base to follow this sort of procedure in the open sea.) Once an adequate model of a reasonably large ocean region (for example the North Atlantic) is achieved, parameterization schemes that allow simulation of the desired larger-scale features with lower resolution can be explored and larger regions on longer time-scales can be studied. Until the fine resolution to allow all important dynamic effects is achieved, and adequate detailed verification data are obtained allowing modellers to work back up to

limited resolution, it may be difficult to devise parameterization schemes
that will allow adequate simulations of the real ocean for large regions and
long time-scales. Work towards these goals will no doubt continue.

There is also the question of whether or not a complex non-linear system, such
as the ocean or the atmosphere or the combined ocean-atmosphere system, has a
unique statistically steady solution. Examples of much simpler non-linear
systems which are always oscillating can be constructed. The decade-to-
decade variations in climate that occur with apparently the same external
solar and tidal forcing suggest that the combined system may be able to exist
in a number of quasi-steady states with jumps from one to another when a
large perturbation occurs. There are variations in the forcing, however, (for
example, variations in the solar wind with the sun spot cycle and the earth's
orbital parameters) so the system may be predictable given the right data.
There is some evidence that very long-term climatic fluctuations are associated with the periodicities of the earth's orbital variations. Much more
work will have to be done before this question of uniqueness can be answered
with any assurance.

Much work has been done on the time-dependent motions, which we have mentioned
only briefly, as our concern has been with the large-scale average circulation. This work starts with considerations of planetary or Rossby waves
mentioned in Chapter 12, and includes such things as effects of topography on
these waves and their interaction with one another as non-linear effects become more important. It leads ultimately, using numerical simulations, to an
investigation of the behaviour of the strongly non-linear case of quasi-two-
dimensional or geostrophic turbulence. The mesoscale eddies, as they are
termed by oceanographers, are probably phenomena of this type. Because these
eddies may interact with the mean flow, this work on time-dependent motions
may be very helpful in our search for a fuller understanding of how the
general circulation works.

Much more information is needed about this kind of turbulence and also the
smaller-scale three-dimensional turbulence which is responsible for frictional
and mixing effects. At present, parameterization of these effects is crude,
largely due to the lack of knowledge of these phenomena in the ocean and how
they interact with the large-scale flow. Further work examining various
turbulence models and how these different turbulence models affect the large-
scale circulation is needed. Then observational programmes need to be con-
ducted to gather data to determine which type of turbulence model is most
representative.

Because of the small spatial scale and the long time-scale, the logistics
of getting good observations of such phenomena as the oceanic eddies are
extremely difficult. Large efforts have been made by groups from the USSR
(the POLYGON Experiment) and the USA-UK (the Mid-Ocean Dynamics Experiment,
or MODE). Each has observed one eddy fairly well and fringes of others.
More of this observational work is underway; POLYMODE, a combination of the
USSR-USA-UK efforts, has begun. As part of these projects a considerable
amount of theoretical work, both analytical and numerical, has been done to
help to design the sensor arrays to optimize the data that have been and are
being collected. In addition to these detailed array studies, examination
of historical bathythermograph (a temperature-depth recorder) data and new
more detailed bathythermograph sections are being examined to get a better
idea of the statistics of the existence and intensity of such eddies in
various ocean regions. These relatively large projects like POLYMODE, in-

volving many scientists and strong interactions between theoretical and observational oceanographers, are likely to be a common method of future oceanographic investigations.  In the past, most oceanographic investigations have involved rather smaller groups of people working independently on partic- ular problems.  While this traditional method will continue to be used, many oceanographic problems will require the much larger project approach to solve them.   These large programmes of direct observations of deep ocean circulation are likely to continue for many years.  The results of theory, both analytical and numerical, must be used to design the observational programme so that data can critically test the theory.  The observational results will no doubt lead to more modifications of our theoretical ideas until we achieve a better understanding of the ocean than we presently have.

Studies of the upper mixed layer of the ocean, mentioned briefly in Chapter 10, will continue.  As noted there, understanding of the development and evolution of this layer, e.g., the formation and breakdown of the seasonal thermocline at mid-latitudes, is important for oceanographic purposes, both physical and biological, and for meteorological purposes.

On the longer time-scale the ocean must play an important role in determining the earth's climate and its variations.  Thus, a better understanding of the ocean general circulation is required if we are to obtain a sufficient under- standing of the physical basis of climate to be able to predict climatic variations.  Because of the effect of climate variations on human activities, particularly agriculture, the study of climate and the role the ocean plays in it has become an active area of research which is likely to continue for many years to come.

The problem of generation of surface gravity waves by the wind continues to be an area of active research.  We still do not have an accurate quantitative understanding of this process.  However, the new work described briefly in Chapter 12 has suggested some interesting new possibilities.  Further investi- gation of these possibilities may in future lead to a better understanding of this process.

While the dynamic oceanography of coastal and estuarine regions is beyond the scope of this book it should be noted that the study of dynamic oceanography in these regions is also an active area of research.  Studies of coastal up- welling regions and estuaries have important applications because they are such biologically important regions.  The need for better navigation and for better knowledge of the transport of materials both natural and man-made are other applications of the study of physical oceanography of coastal regions.

Attempts to calculate the tidal response of the oceans from the basic equations of motion are continuing.  Such attempts for the deep ocean are being aided by the data obtained with deep-sea internally-recording tide gauges.  Attempts at calculating tides and more particularly tidal currents are also being made in coastal areas.  While we can always make tidal predictions empirically as described in Chapter 13, modelling which allows prediction of tidal currents is very important because measurement of these currents in detail is difficult, time-consuming and expensive.

Development of new tools to assist the dynamic oceanographer in his investi- gations is also continuing.  The continued development of better moored current meters is an example.  They will provide longer term measurements more reliably.  The data provided by satellites is another example.  Sea

surface temperatures and infrared photographs obtained from satellites help
to show the paths of ocean currents.  The technique of obtaining sea surface
elevations with a relative accuracy of about 10 cm may soon be available.
Such observations would give information on the tides and, with suitable
averaging, information about the stronger currents of the general circulation.
If an accuracy of the order of 1 cm could be achieved then such observations
along with geostrophic calculations would provide a powerful tool for determining the general ocean circulation for comparison with theory.

The study of fine structure (vertical variations with scales of one to a few
metres) and of microstructure (scales of centimetres to millimetres, i.e.,
the small-scale turbulence) is another area of active research.  Such studies
should eventually lead to a better description and understanding of the
details of the mixing and friction processes and better parameterizations for
vertical eddy viscosity and diffusivity.  Better parameterization for the
horizontal eddy processes for small scales should result too.  For large
scales, such parameterization may well come from the study of mesoscale
eddies.

The study of the Antarctic region and the Circumpolar Current which we
mentioned briefly in Chapter II has been receiving renewed attention in the
International Southern Ocean Study (ISOS) which is part of the International
Decade of Ocean Exploration (IDOE) sponsored by the United States of America
National Science Foundation.  (The Coastal Upwelling Ecosystems Analysis
programme (CUEA) is another example of a large scale programme with physical
oceanographic aspects.  MODE was and POLYMODE is also partially supported by
IDOE.)  The Antarctic Circumpolar Current is the main feature of the only
ocean circulation system which has no complete meridional boundary, and is
therefore extremely interesting.  While it is agreed that wind-driving is
very important, there is a very wide range of transport estimates for the
current and controversy over the importance of other possible driving mechanisms.  With the more extensive moored current meter data being collected in
ISOS as well as other observations, a better description and understanding
of this circulation system should be achieved in the near future.

Clearly, our knowledge of the sea is incomplete at present and much remains
to be done.  Thus for the student interested in observing and interpreting
the oceans there are still many opportunities.

# APPENDIX 1

# Mathematical Review with Some
# Elementary Fluid Mechanics

The purpose of this Appendix is to provide a brief review of some symbolism, mathematical procedures and some aspects of fluid mechanics relevant to the text, primarily for students whose field is not in the physical sciences.

The main symbols used to represent physical quantities in equations in this text have been given at the beginning of the book (List of Symbols).

For coordinate axes we use a right-handed system (Fig. A.1a) with the positive x-axis directed horizontally to the east, the positive y-axis horizontally to the north, and the positive z-axis vertically upward. The corresponding velocity components are then u positive to the east, v positive to the north and w positive upward. (It should be noted that *The Oceans* (Sverdrup et al., 1946) uses a left-handed system with the z-axis positive downward.)

## SCALARS AND VECTORS

*Scalar* quantities are those which are expressed by a number and a unit only (e.g., temperature, mass) while *vector* quantities possess direction in addition (e.g., velocity, acceleration, force). The underlining of a symbol is used to indicate that it is a vector, e.g., $\underline{V}$. If we use the same symbol without the underline it means the magnitude or size of the vector quantity regardless of direction. [For velocity (V), the magnitude (V) is called the 'speed'. In some texts 'velocity' is used where 'speed' would be appropriate; we have tried not to do so.] It is often convenient to split a vector quantity into components, e.g., in Fig. A.1b the vector $\underline{V}$, which is directed at an angle $\theta$ to the x-axis can be resolved into components: $V \cdot \cos\theta = u$ to the east and $V \cdot \sin\theta = v$ to the north. When we wish to be specific about direction we write $V \cdot \cos\theta$ as $\underline{i} \cdot u$ where $\underline{i}$ is a 'unit vector' directed to the east (positive x direction). Then $-\underline{i} \cdot u$ would represent a component u directed to the west (negative x direction). Similarly $\underline{j}$ and $\underline{k}$ are unit vectors in the north (+y) and upward (+z) directions respectively. Another example (Fig. A.1c) is where the acceleration due to gravity g, which acts vertically, is resolved into a component $g \cdot \sin\theta$ down the sloping surface which is at angle $\theta$ to the horizontal and a component $g \cdot \cos\theta$ perpendicular to the surface. In oceanographic applications of this resolution the angle $\theta$ is generally small so that $g \cdot \sin\theta$ is a small quantity while $g \cdot \cos\theta$ is large (nearly equal to g).

214

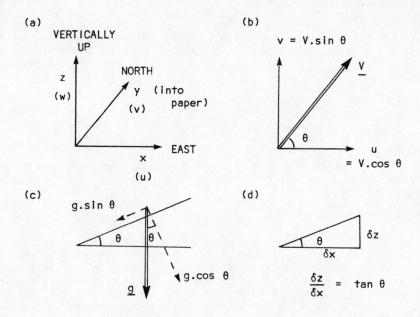

Fig. A.1.    (a) Orientation of axes used in this text with corresponding
             velocity components (u, v, w), (b) rectangular components of a
             vector (V), (c) components of g on a slope, (d) gradient $\delta z/\delta x$)
             of a slope.


## DERIVATIVES

The abbreviation $\delta z$ means 'a small distance in the z direction' and $\delta x$ means
'a small distance in the x direction'.  Then the quotient $\delta z/\delta x$ is physically
a measure of the slope of the surface (e.g., Fig. A.1d) relative to the horizontal, i.e., $\delta z/\delta x = \tan\theta$.  In the limit, when $\delta x$ is assumed to approach
zero ($\delta x \to 0$) so that we are looking at a point in space, $\delta z/\delta x$ is written as
$\partial z/\partial x$ which is called the 'derivative', 'gradient' or 'rate of change' of z
with respect to x.  Examples of other derivatives common in physical oceano-
graphy are $\partial S/\partial x$, $\partial T/\partial x$, $\partial S/\partial z$, $\partial p/\partial z$ etc.  When the derivative is written
with $\partial$, e.g., $\partial S/\partial x$, it implies that the quantity S is known to vary with x,
y, z and t [i.e., $S = S(x,y,z,t)$] but in this case we are only interested in
the change with respect to x and are assuming that y, z and t remain fixed.

One special case is when the derivative $\partial/\partial t$ of a quantity is zero, e.g.,
$\partial S/\partial t = 0$, $\partial u/\partial t = 0$.  There is no change of the quantity with time, but it
does not imply that the quantity itself is zero, e.g., $\partial u/\partial t = 0$ means that
there is no change of u with respect to time at any point in the region under
study but u need not be zero, i.e., the water may be moving but at a constant
(steady) speed at each point.  This situation is referred to as the *steady
state*.  Note also that the combined statements $S = S(x,y,z)$, $\partial S/\partial t = 0$ imply
that S does not change with time anywhere but the value of S may differ from
one point to another.

## The Individual or Total Derivative

In fluid dynamics a special case arises when a quantity q varies with position
$(x,y,z)$ and with time $(t)$, i.e., $q = q(x,y,z,t)$. Then we will show that the
time derivative $dq/dt$, called the 'individual' or 'total' derivative (because
it is the time change of q following a particular piece of the fluid) is given
by:

$$\frac{dq}{dt} = \frac{\partial q}{\partial t} + u \cdot \frac{\partial q}{\partial x} + v \cdot \frac{\partial q}{\partial y} + w \cdot \frac{\partial q}{\partial z} \, . \qquad (A.I)$$

$$\underset{\text{term}}{\text{Local}} \qquad \underset{\text{terms}}{\text{Advective}}$$

Physically this equation states that q may vary with time $(\partial q/\partial t)$ at a position $(x,y,z)$ and also vary as the fluid moves from this point to another point
$(x + \delta x, y + \delta y, z + \delta z)$. The first term on the right of equation A.I is
then called the 'local' term and the other three are the 'advective' terms
because they are related to the flow (advection) components u, v and w.

To derive this expression, consider first a case in which the value of q at
all points does not change with time - the steady state. Mathematically we
write $\partial q/\partial t = 0$. However, q may still change with position. In this case,
a small 'parcel' of fluid or 'fluid element' moving through the field must
undergo changes, i.e., $dq/dt \neq 0$ unless q is the same everywhere. Initially
we will suppose that there is motion only in the x direction and variations
only in the x direction so that at time t a particle is at point x with
property $q(x)$ while at a slightly later time $(t + \delta t)$ it is at $(x + \delta x)$ with
property $q(x + \delta x)$. Now using Taylor's series expansion we can write:

$$q(x + \delta x) = q(x) + (\partial q/\partial x) \cdot \delta x + \text{terms of order (proportional to)}$$
$$(\delta x)^2 \text{ or smaller}$$

which can be neglected (since in the limit as $\delta x \to 0$ these terms will vanish).
The property change from x to $(x + \delta x)$ is therefore $(\partial q/\partial x) \cdot \delta x$ and the rate
of change following the motion is:

$$\frac{\text{property change}}{\text{time change}} = \frac{\left(\frac{\partial q}{\partial x}\right) \cdot \delta x}{\delta t} = \frac{\partial q}{\partial x} \cdot \frac{\delta x}{\delta t} \, .$$

In the limit, as $\delta t \to 0$, $\delta x/\delta t \to u$, and the rate of change $= u \cdot \partial q/\partial x$ in the
x direction. In the more general case when there are also v and w components
of velocity and variations in all three component directions, the rate of
change associated with the motion of the fluid is

$$u \cdot \frac{\partial q}{\partial x} + v \cdot \frac{\partial q}{\partial y} + w \cdot \frac{\partial q}{\partial z} \, .$$

This is the 'advective' component of the property change.

If we now include changes with time at the point itself in the fluid, i.e.,
$\partial q/\partial t$ (the 'local' rate of change) the 'total' derivative will be:

$$\frac{dq}{dt} = \frac{\partial q}{\partial t} + u \cdot \frac{\partial q}{\partial x} + v \cdot \frac{\partial q}{\partial y} + w \cdot \frac{\partial q}{\partial z} \, .$$

Fig. A.2.   (a) Significance of an integral, (b) area under curve obtained
            by summation of areas of strips.


The reason for the alternative name 'individual' derivative is because it is
the derivative following the motion of an individual particle of the fluid.

The quantity q may be a scalar property of the fluid, e.g., salinity (S) or
temperature (T), or it may be a vector quantity, e.g., velocity (V). In the
latter case, there will in general be three components to the individual
derivative as:

$$\text{x direction:} \quad \frac{du}{dt} = \frac{\partial u}{\partial t} + u \cdot \frac{\partial u}{\partial x} + v \cdot \frac{\partial u}{\partial y} + w \cdot \frac{\partial u}{\partial z} \,,$$

$$\text{y direction:} \quad \frac{dv}{dt} = \frac{\partial v}{\partial t} + u \cdot \frac{\partial v}{\partial x} + v \cdot \frac{\partial v}{\partial y} + w \cdot \frac{\partial v}{\partial z} \,,$$

$$\text{z direction:} \quad \frac{dw}{dt} = \frac{\partial w}{\partial t} + u \cdot \frac{\partial w}{\partial x} + v \cdot \frac{\partial w}{\partial y} + w \cdot \frac{\partial w}{\partial z} \,.$$


## INTEGRALS

Integration essentially means summation. For instance, in Fig. A.2a we may
want to determine the value of the single-shaded area between the curve, the
x axis and the vertical lines at $x = x_1$ and $x = x_2$. We can do so by dividing
the area into narrow strips, such as the double shaded one whose height is $y_1$
and width $\delta x$, and whose area is therefore $\delta A = y_1 \cdot \delta x$ and then summing the

areas of all the strips so that the total area

$$A = \Sigma \delta A = \int_{x_1}^{x_2} y \cdot \delta x \quad ,$$

and then summing the areas of all the strips so that the total area

$$A = \int_{x_1}^{x_2} y \cdot dz \quad .$$

If y is a simple function of x there are standard rules for 'integrating y with respect to x', i.e., determining the value of the integral, but if y is a complicated or irregular function of x, as in Fig. A.2b, there may be no known rule. It will then be necessary to go back to the basic procedure $A = \Sigma_{x_1}^{x_2} y \cdot \delta x$ by dividing the total area into a number of thin strips and adding the individual areas so that $A = \delta A_1 + \delta A_2 + \delta A_3$ etc. $= y_1 \cdot \delta x + y_2 \cdot \delta x + y_3 \cdot \delta x +$ etc. This procedure is used in Chapter 8 in geostrophic current calculations.

### FIELDS

Physicists use the word *field* for the distribution of a quantity in space, e.g., geographical distributions of temperature or salinity are scalar fields while current distribution patterns represent vector fields, e.g., Fig. A.3b. It must be noted that in such field patterns, isopleths (lines of equal value) of the property generally cannot cross or touch because this would mean that the quantity had two different values at the same point.

### DESCRIPTIONS OF FLUID FLOW

There are two ways of describing the pattern of flow in a region: *Lagrangian* in which we describe or plot the path (trajectory) followed by each fluid particle, specifying when each particle reaches each point in its path, and *Eulerian* in which we describe the velocity (speed and direction) of the fluid at every point in the fluid at every instant of time.

These are illustrated graphically in Fig. A.3 using the North Pacific Ocean as an example. If the circulation does not change with time, then these two figures will be related very simply - each arrow of the Eulerian pattern will be a tangent to a continuous line of the Lagrangian pattern. Therefore, if current measurements are made at a series of points to obtain initially an Eulerian picture, the Lagrangian one may be constructed by drawing continuous lines through the arrows so that the latter are tangents to the lines. In this case the Lagrangian lines will also be true *streamlines* (lines which at any given time are everywhere tangent to the Eulerian velocity). However, if the currents are changing with time, then the streamline pattern, which is an instantaneous feature, will be continually changing and the Lagrangian pattern, which shows the history of the flow, will be a step-by-step summation of the Eulerian pattern and will not be the same as the streamline pattern for any time.

It may be noted that the North Pacific example which we have chosen, the main (long-time average) character of the clockwise circulation probably does not

Fig. A.3.   Flow pattern described in (a) Lagrangian manner (flow lines),
            (b) Eulerian manner (velocities at points).

change much, and the difference between the Lagrangian trajectories and
individual streamline patterns may not be large.  However, in other places,
such as the Western Coral Sea or the Equatorial Indian Ocean, the direction
of the currents reverses during its annual cycle and there is a considerable
difference between streamline and trajectory patterns.

Fig. A.4.   Flow patterns at convergence and divergence at (a,b) a point,
            (c,d) a line, (continued)

The Lagrangian method is commonly used for descriptions of ocean circulation,
using separate patterns for the different seasons if there is a marked seasonal variation.  It is the form in which information comes from measurements
made by following drifting floats or buoys.  However, it is difficult to
handle mathematically.  The Eulerian description is obtained by putting current
meters at a number of fixed points and recording speed and direction at every
point simultaneously.  It is more tractable mathematically and we use it when
investigating the equations of motion in the sea.  In principle one can convert from one to the other; in practice one rarely has enough observations to
do so confidently.


## CONVERGENCES AND DIVERGENCES

A common feature of flow patterns at the sea surface is one where the flow
*converges* toward a point or line (Fig. A.4a,c) or *diverges* from a point or
line (Fig. A.4b,d).  In the convergent patterns, because the water cannot just
disappear it must flow downward at the point or line of convergence.  For the
divergences, the water must come up from below the surface and then flow outward.  These patterns are mathematical idealizations because the down or up
flow has infinitesimal cross-section, which is not possible in the real world
where water is, to a first approximation, incompressible and where the flow
is not just in a mathematically thin surface layer but actually has a significant depth.  In reality, convergence or divergence generally occurs over an
area as shown in Figs. A.4e,f.  These are plan views of the surface with the
arrow lengths representing the flow in the convergent or divergent upper
layer.  Figs. A.4g,h show vertical cross-sections of the flows above them.
Note that Figs. A.4e,f, which show an integrated upper layer volume flow,

PLAN VIEWS OF UPPER LAYER

VERTICAL SECTIONS
OF UPPER LAYER

Fig. A.4 (continued)  Flow patterns at convergence and divergence for (e,f)
         area  - plan view, (g,h) area - side view.

are an Eulerian type of representation while Figs. A.4 g,h are a Lagrangian
form.

Another point to note is that for a convergence or divergence along a line,
it is not essential that the flow be in opposite directions at either side of
the line.  For instance, a flow from the left might meet a stationary body of
water on the right (e.g., a river flowing out into the sea), or a faster
moving body of water on the left might be overtaking a slower moving water
body on the right.  Also the convergence area may be moving, not stationary.

It is often easy to identify convergences in the sea because objects floating
at the surface are concentrated close to the centre or line of convergence.
Divergences are less commonly rendered visible but in coastal waters where
there are strong tidal currents, the subsurface water is sometimes forced to
the surface by bottom irregularities and 'boils' of water come up to the

$$p_z = - \rho.g.z \qquad\qquad p_z = - \sum_1^3 \rho_n.g.z_n \qquad\qquad p_z = - \int_0^z \rho.g.dz$$

Fig. A.5.    Pressure in a fluid: (a) density ($\rho$) constant, (b) density ($\rho_n$)
varying in steps, (c) density ($\rho_z$) varying smoothly with depth.

surface and flow outward, often too rapidly to row against.  The surfaces of
these boils are usually smoother than the surrounding water and are visible
for this reason.

### HYDROSTATIC PRESSURE

The total pressure at a point below the water surface is equal to the sum of
the atmospheric pressure at the surface and the pressure due to the weight
of fluid above the point.  In most dynamic oceanography calculations it is
usual to neglect the atmospheric pressure term and use only the pressure due
to the weight of fluid, the *hydrostatic pressure* term.  The rationale for
this procedure is that the total atmospheric pressure is equivalent to the
hydrostatic pressure due to a column of about 10 m of water, while the normal
variations of atmospheric pressure are equivalent to only about ±0.3 m of
water pressure, which is negligible for most purposes.

Referring to Fig. A.5a if the water is uniform in density $\rho$ (ignoring the
compressibility effect) the hydrostatic pressure at level z (depth h = -z)
is $p_z = - \rho \cdot g \cdot z$ where g is the acceleration due to gravity which is here
assumed to be independent of depth.  This simple situation is rarely found in
the sea where the density generally increases with depth.  If the water column
were made up, for example, of three layers each of uniform density as in
Fig. A.5b then the pressure at $z = z_1 + z_2 + z_3$ would be $p_z = -(\rho_1 \cdot z_1 + \rho_2 \cdot z_2 +$
$\rho_3 \cdot z_3) \cdot g$.   More generally in the ocean, the density increases with depth in
the manner shown at the right in Figure A.5c with an upper mixed layer of
nearly uniform density, then a zone of increasing density (the 'pycnocline'
zone) grading into a deeper zone of more slowly increasing density.  In this
case, if we write $dp = -\rho_z \cdot g \cdot dz$ for the pressure due to a small layer of
water thickness dz where the density is $\rho_z$, then the total hydrostatic

Fig. A.6.    Forces related to a sloping surface.

pressure at level z will be $p_z = \int_0^z -\rho_z \cdot g \cdot dz$ .    The minus signs occur
because z increases upward; with z = 0 at the surface, z is negative below
the surface while $p_z$ must be positive.  If the density happens to vary with
depth according to some mathematical function which can be integrated, then
this integral may be evaluated directly.  However, it is usually necessary
to use the procedure of Fig. A.5b, dividing the total water column into a
sufficient number of thin layers, each of essentially uniform density, to
represent closely the real distribution with depth.  For analytical purposes,
the shape of density variation of Fig. A.5c has sometimes been represented by
an upper mixed-layer of depth h of uniform density $\rho_1$ and a lower layer of
density $\rho_2 = \rho_d - (\rho_d - \rho_1) \cdot \exp(1 + z/h)$ for z < -h, so that in the lower layer
the density approaches asymptotically the deep water value $\rho_d$.

### SLOPE EFFECTS

If a block of wood (mass, m) were placed on a sloping surface (Fig. A.6a) it
is quite likely that it would remain stationary because the component
$mg \cdot \sin \theta$ of its weight (mg) down the slope would be less than the possible
friction force up the slope.  However, if *water* were poured on the slope
(Fig. A.6b), the frictional force between it and the slope would be much
smaller than the weight component down the slope initially since there is no
friction until motion begins; the water would flow down the slope with increasing speed until the friction associated with the flow balanced the weight
component down slope.  If we had a container of water and established a slope
on the surface (Fig. A.6c) not only would the surface water move to the left,

but the whole body of water would tend to move to the left (assuming that the water were uniform in density). The reason is that at any level in the water, the pressure $p_R$ on the right side of a small cube of water (Fig. A.6d) would be greater than that on the left side $p_L$ because of the greater depth on the right. Therefore there would be a net force to the left and a consequent tendency for the cube to move to the left. For a fluid of uniform density this argument applies at all depths from the surface to the bottom. It is an example of a *barotropic* force system, in which the isobaric (equal pressure) surfaces are all parallel to each other (and to the surface).

The above statements apply to the situation when the water is initially stationary. It is shown in the text that if the water is moving appropriately (into the paper in the northern hemisphere in this case) the Coriolis force (see Chapter 8) to the right may be sufficient to balance the hydrostatic pressure force to the left and the slope may be maintained.

A situation in which the net horizontal hydrostatic force does not penetrate to the bottom could occur if the density of the water in the upper layer above level AA' in Fig. A.6c decreased progressively to the right so that the hydrostatic pressure $p_A = -\int_0^{z_A} \rho \cdot g \cdot dz$ remained constant along AA'. Then there would still be a tendency for the water above AA' to move to the left but if the density were uniform below AA' there would be no resultant hydrostatic pressure force and no tendency for the deep water to move. In the layer above AA', where the density varies with position, the isobaric surfaces will be inclined to each other, rather than parallel, and the force system is then described as *baroclinic*.


COMPRESSIBILITY

The compressibility K of a fluid is defined as $K = -\frac{1}{V} \cdot \frac{dV}{dp}$ where V is volume and p is pressure, and for a fluid to be incompressible K = 0. Now we can write

$$K = -\frac{1}{V} \cdot \frac{dV}{dp} = -\left(\frac{1}{V} \cdot \frac{dV}{dt}\right) \cdot \left(\frac{dt}{dp}\right) = -\left(\frac{1}{V} \cdot \frac{dV}{dt}\right) / \left(\frac{dp}{dt}\right)$$

where dV/dt and dp/dt are changes with time of volume and pressure respectively, following a small parcel (in the limit infinitesimal) of fluid. If the pressure is changing but K = 0, we conclude that $\frac{1}{V} \cdot \frac{dV}{dt} = 0$. Now for a constant mass m of fluid of volume V and hence density $\rho = m/V$:

$$\frac{1}{\rho} \cdot \frac{d\rho}{dt} = \frac{V}{m} \cdot \frac{d}{dt}\left(\frac{m}{V}\right) = -\frac{1}{V} \cdot \frac{dV}{dt}$$

so that we can regard incompressibility as meaning either $\frac{1}{V} \cdot \frac{dV}{dt} = 0$, or

$$\frac{1}{\rho} \cdot \frac{d\rho}{dt} = 0 \quad .$$

### CENTRIPETAL AND CENTRIFUGAL FORCES

Acceleration is defined as the rate of change of the velocity of a body and
may consist of a rate of change of the speed and/or of the direction.  If the
body has mass, a resultant force along the direction of motion is required to
cause the change of speed.  If the direction of motion is changing, the body
must be travelling on a curved path and there must be another resultant force
directed inward toward the centre of curvature of the path.  This force, which
acts on the body itself, is called the *centripetal* force.  Now Newton's Third
Law of Motion states that to every acting force there must be a reacting
force, equal in magnitude, opposite in direction and *acting on another body*.
This reaction is called the *centrifugal* force and it is important to note
that it does not act on the body itself, but on something else.

A simple example of this force pair is provided when a small mass attached
to a string is whirled round in a circle  the inner end of the string being
held in the hand.  Then the inward force of the string provides the *centripetal force on the mass,* while the outward pull of the string *on the hand* is
the centrifugal force.

Some authors study the dynamics of such a system by using a fictional outward
force on the mass (equal to its mass × its acceleration) and call this a
'centrifugal force'.  While the procedure of using an outward 'mass acceleration' is a legitimate (and often convenient) device for solving the dynamics,
it is incorrect to call this the 'centrifugal' force.  We prefer to treat
problems in rotation in terms of the *inward* directed centripetal force on the
mass itself.

In the example above, the tension in the string acting on the rotating mass
provides the centripetal force while in the astronomical case, the gravitational attraction of the earth on the moon, for example, provides the centripetal
force needed to maintain the moon in orbit.  Likewise, the reciprocal
gravitational attraction of the moon on the earth provides the centripetal
force needed to make the earth revolve about the centre of mass of the earth-
moon system.

# APPENDIX 2
# Units used in Physical Oceanography

## INTRODUCTION

Hitherto, physical oceanographers have used a mixed system of units which has its advantages for those thoroughly familiar with it but which offers some difficulties for the beginner in this field. Because the International System of Units (abbreviated SI) is now coming into general use we have used it in this text. In SI there are base units and derived units (from the base units), and a number of temporary units are accepted but will be phased out of use eventually. In the following we will first list the base, derived and temporary units used in dynamic oceanography and then will relate them to the older mixed units. Our basic reference for SI practice is the Metric Practice Guide published by the Standards Council of Canada (reference CAN-3-001-02-73, CSA Z234.1-1973).

## BASE UNITS

The base units used in this text, their abbreviations and physical dimensions are:

| Quantity | Base Unit | Abbreviation | Dimension |
|----------|-----------|--------------|-----------|
| Length | – metre | m | [L] |
| Mass | – kilogram | kg | [M] |
| Time | – second | s | [T] |
| Thermodynamic temperature | – Kelvin | K | [K] |

## DERIVED AND TEMPORARY UNITS

The units are given with their abbreviations and under the name of the unit is given the symbol used in equations in this text. On the right are given the physical dimensions of the unit in terms of length (L), mass (M), time (T) and thermodynamic temperature (above absolute zero) (K).

| | | | | |
|---|---|---|---|---|
| Length:<br>(L) | 1 centimetre | (cm) | = $10^{-2}$ m | [L] |
| | 1 decimetre | (dm) | $10^{-1}$ m | |
| | 1 kilometre | (km) | = $10^{3}$ m | |
| | 1 International nautical mile (n ml) = 1,852.0 m (temporary unit) | | | |
| Mass:<br>(m,M) | 1 gram | (g) | = $10^{-3}$ kg | [M] |
| | 1 tonne | (t) | = $10^{3}$ kg | |

| Time: | 1 minute | (min) | = | 60 s | $[T]$ |
|---|---|---|---|---|---|
| $(t,T)$ | 1 hour | (h) | = | 60 min | |
| | 1 day | (d) | = | 24 h = 86,400 s (mean solar day) | |
| | | | | (86,164 s = 1 sidereal day) | |
| | 1 year | (a) | = | not defined in SI, here taken as | |
| | | | | 365 d. | |

Area:            1 $m^2$                                            $[L^2]$
(A)

Volume:          1 $m^3$                                            $[L^3]$
(V)
                 1 litre (ℓ) = 1 $dm^3$ = $10^{-3} m^3$
                 (not used for high precision measurements, not SI)

Speed; components:   1 $m\,s^{-1}$ = $10^2\,cm\,s^{-1}$                    $[LT^{-1}]$
(V; u,v,w; C)        1 knot (kn) = 1 naut ml $h^{-1}$ = $0.514\,m\,s^{-1}$
                       (temp. unit)

Acceleration:    1 $m\,s^{-2}$                                       $[LT^{-2}]$
(a)

Density:         1 $kg\,m^{-3}$ = $10^{-3}\,g\,cm^{-3}$                    $[ML^{-3}]$
(ρ)

Relative density:   Value given relative to water for liquids
(d)                 and to air for gases - formerly 'specific
                    gravity' (pure number - no units)

Specific volume:    1 $m^3\,kg^{-1}$ = $10^3\,cm^3\,g^{-1}$                 $[M^{-1}L^3]$
(α)

Force:           1 Newton (N) = 1 $kg\,m\,s^{-2}$ = force required   $[MLT^{-2}]$
(F)                to give 1 kg mass an acceleration of
                   1 $m\,s^{-2}$ (= $10^5$ dynes - not in SI).

Pressure:        1 Pascal (Pa) = 1 $N\,m^{-2}$ (= 10 $dyn\,cm^{-2}$)    $[ML^{-1}T^{-2}]$
(p)

Energy:          1 Joule (J) = 1 $N \cdot m$                          $[ML^2T^{-2}]$
(E, W)

Dynamic Viscosity:   1 Pascal second (Pa s) = 10 poise (P)          $[ML^{-1}T^{-1}]$
(μ)                                    (not in SI)

Kinematic viscosity:   1 $m^2\,s^{-1}$ = $10^4\,cm^2\,s^{-1}$ = $10^4$ Stokes (St)   $[L^2T^{-1}]$
$(\nu = \mu\rho^{-1})$                                  (not in SI)

Kinematic diffusivity: 1 $m^2\,s^{-1}$                               $[L^2T^{-1}]$
(κ)

Temperature:     The Celsius temperature (°C) is the differ-     $[K]$
(T)              ence between the thermodynamic temperature
                 T and the temperature $T_0$ = 273.15 K
                 (unit = 1 K)

## UNITS USED IN DYNAMIC OCEANOGRAPHY AND SOME NUMERICAL VALUES

Some of these units are peculiar to physical oceanography; numerical values
for some of the quantities are given for illustration.

Sverdrup :   a unit of volume flow = $10^6$ $m^3$ $s^{-1}$                                    $[L^3T^{-1}]$

Salinity :   the physical definition is: 'The total quantity of solid
(S)          material in grams contained in one kilogram of sea water when
             all the carbonate has been converted to oxide, the bromine and
             iodine replaced by chlorine and all organic matter completely
             oxidised.' It is expressed as parts per thousand (‰) and
             has no units. The mean value for the oceans is about 34.7 ‰,
             with values in the Red Sea up to 40 ‰ and values close to
             0 ‰ in coastal regions near rivers. In practice, salinity
             is now determined by measuring the electrical conductivity
             relative to a standard sea water and converting to salinity
             using tables prepared from laboratory determinations of the
             conductivity/salinity relationship.                  [pure number]

Density:     Is a function of S, T and p, i.e., $\rho = \rho(S,T,p)$.        $[ML^{-3}]$
(ρ)          For sea water of salinity 35.00 ‰, temperature 10.00 °C at
             standard atmospheric pressure (= zero hydrostatic pressure),
             $\rho_{35,10,0}$ = 1,026.97 kg $m^{-3}$. Strictly speaking, the quantity
             which the oceanographer calls 'density' is operationally
             measured relative to pure water as the standard and is really
             'relative density'. However, in physical equations it must be
             treated dimensionally as density.

Sigma-t:     This quantity is introduced for convenience and is de-    $[ML^{-3}]$
($\sigma_t$)          fined as: $\sigma_t = (\rho_{S,T,0} - 1,000.00)$. For the sample of sea
             water above, the value of $\sigma_t$ = 26.97. (It is usual to omit the
             units (kg $m^{-3}$) when stating values for $\sigma_t$, which is usually used
             for descriptive purposes rather than as a component in equations.)

             In the mixed unit system, with $\rho$ in g $cm^{-3}$, sigma-t is defined
             as $\sigma_t = (\rho - 1) \times 10^3$, which gives the same numerical value as
             the SI unit definition given above.

             Note that although sea water density may be stated to six
             significant figures, and $\sigma_t$ to four figures, the absolute values
             are not known to this accuracy. Differences between densities
             for typical oceanic waters may be accurate to the second decimal
             place (0.01 kg $m^{-3}$) but absolute values only to the first
             decimal place (0.1 kg $m^{-3}$), i.e., five significant figures in $\rho$
             or three in $\sigma_t$.

Specific volume:  Is a function of S, T and p, i.e., $\alpha(S,T,p)$.        $[M^{-1}L^3]$
($\alpha = 1/\rho$)   For sea water of salinity 35.00 ‰, temperature 10.00°C
             at standard atmospheric pressure, $\alpha_{35,10,0}$ = 0.973738 × $10^{-3}$ $m^3 kg^{-1}$.
             Note that in the symbol $\alpha_{35,10,0}$ the zero for p indicates that
             the value is for zero hydrostatic pressure (but at atmospheric
             pressure as would be the case in a laboratory determination).

Specific volume anomaly:  Defined as $\delta = \alpha_{S,T,p} - \alpha_{35,0,p}$        $[M^{-1}L^3]$
($\delta$)              with units $m^3$ $kg^{-1}$.

Thermosteric anomaly:  A term in the expansion of $\delta$ (see text,          $[M^{-1}L^3]$
$(\Delta_{S,T})$           Chapter 2).  SI units are $m^3$ $kg^{-1}$; for $\sigma_t$ = 26.97, the value
of $\Delta_{S,T}$ = 109.8 $\times$ $10^{-8}$ $m^3$ $kg^{-1}$.  In the old mixed units system
the corresponding value is 109.8 $\times$ $10^{-5}$ $cm^3$ $g^{-1}$, or 109.8 cl
(centilitres) $tonne^{-1}$ to avoid having to write the power of 10.
(Here 1 tonne = 1,000 kg, 1 cl $t^{-1}$ = $10^{-8}$ $m^3$ $kg^{-1}$).

Pressure:      1 bar = $10^3$ mb (= $10^6$ dyn $cm^{-2}$) = $10^5$ Pa = 100 kPa,        $[ML^{-1}T^{-2}]$
(p)            (not SI)        1 decibar = 1 db = $10^4$ Pa =  10 kPa.
Standard atmospheric pressure = 1,013.25 mb = 101.325 kPa.
In the open ocean, the pressure at a geometrical depth of
1,000 m is about 1,010 db = 10,100 kPa.

Geopotential:  The work done per unit mass to raise a body vertic- $[L^2T^{-2}]$
($\Phi$)             ally through a small distance z in the vicinity of the
earth = g $\cdot$ z Joules $kg^{-1}$.  In SI the unit of $\Phi$ is 1 J $kg^{-1}$ = 1 $m^2$ $s^{-2}$,
and the acceleration due to gravity g = 9.80 m $s^{-2}$, then for a
vertical lift of 1 m, $\delta\Phi$ = 9.80 J $kg^{-1}$.

In the past in dynamic oceanography, a quantity called 'dynamic
height' (D) has been used.  This is geopotential expressed in
units such that 1 dynamic metre = 10.0 J $kg^{-1}$ = 10.0 $m^2$ $s^{-2}$.
Then the dynamic height difference between the sea surface and
1,000 m geometrical depth = -980 dyn m for g = 9.80 m $s^{-2}$.

Viscosity:     For sea water of S = 35‰ at T = 10°C,
($\mu$, $\nu$,
$A_x, A_y, A_z$)    Molecular viscosity – Dynamic   : $\mu \simeq 1.4 \times 10^{-3}$ kg $m^{-1}$ $s^{-1}$
– Kinematic : $\nu = \mu \cdot \rho^{-1} \simeq 1.4 \times 10^{-6}$ $m^2$ $s^{-1}$ .

Eddy viscosity        – Kinematic : $A_x$, $A_y$ up to $10^5$ $m^2$ $s^{-1}$,
$A_z$   up to $10^{-1}$ $m^2$ $s^{-1}$.

Diffusivity :  For sea water, kinematic:
($\kappa_T$, $\kappa_S$)    Molecular diffusivity:  for heat, $\kappa_T \sim 1 \times 10^{-7}$ $m^2$ $s^{-1}$
for salt, $\kappa \sim 1 \times 10^{-9}$ $m^2$ $s^{-1}$.

Eddy diffusivity:  same ranges as eddy viscosity above.

# Suggestions for Further Reading and for Reference

<u>TEXTS</u>

Adams, W.A. (Ed.); *Tsunamis in the Pacific Ocean*, East-West Centre Press, Honolulu, 1970, p. 513. A number of research papers on tsunami problems.

Barber, N.F.; *Water waves*, Wykeham Publ., 1969, p. 142. Descriptions of wave characteristics and behaviour with only a little mathematics. Intended for high school students.

Bascom, W.; *Waves and Beaches*, Anchor, Doubleday, 1964, p. 267. A lively account of experimental studies of ocean waves, particularly near the shore, and of their relations to beach form.

Batchelor, G.K.; *Introduction to fluid dynamics*, Cambridge University Press, 1967, p. 615. A review of the field; graduate level.

Bjerknes, V. *et al.*; *Physikalische Hydrodynamik*, Springer, Berlin, 1933, p. 797. For the transformations from fixed to rotating axes from the equations of motion, and an immense quantity of other material in physical hydrodynamics.

Darwin, G.H.; *The tides and kindred phenomena in the solar system*, Houghton, Mifflin, 1911; reprint Freeman, San Francisco, 1962, p. 378. The text of a series of public lectures with excellent non-mathematical discussions/expositions of topics in this field.

Defant, A.; *Ebb and Flow, the Tides of Earth, Atmosphere and Water*, University of Michigan Press, 1958, p. 121. A very interesting descriptive account with illustrations.

Defant, A.; *Physical Oceanography*, Pergamon Press, 1960, p. 1319. An advanced level text. Volume I, Pt. 1, is descriptive while Pt. 2 and all of Vol. II are dynamical.

Dietrich, G. and K. Kalle; *General Oceanography - an Introduction*, Wiley, 1963, p. 588. A fairly comprehensive text on descriptive and dynamical oceanography, relatively little mathematics, good illustrations.

Dyer, K.R.; *Estuaries - a Physical Introduction*, 1973, p. 140. A summary of the descriptive and dynamic oceanography of estuaries.

Fomin, L.M.; *The Dynamic Method in Oceanography*, Elsevier, 1964, p. 212. Devoted to the use of the geostrophic method and dynamic topographies for current determination.

Godin, G.; *The analysis of tides*, Univ. of Toronto Press, 1972, p. 264. A detailed, advanced description of the various techniques.

Goldberg, E.D. *et al.* (Eds.); *The Sea: Ideas and Observations,* Wiley-
    Interscience, Vol. 6, 1977, p. 1048.  Marine Modeling.  A collection of
    papers on modeling of physical, geological, chemical and biological
    systems in the sea.

Hill, M.N. (Ed.);  *The Sea: Ideas and Observations,* Wiley-Interscience,
    Vol. 1, 1962, p. 864.  *Physical Oceanography.*  A collection of advanced
    papers on dynamical oceanography.
    Vol. 2, 1963, p. 554.  *The Composition of Sea Water; Comparative and
    Descriptive Oceanography.*  A series of mainly descriptive papers on the
    chemistry, biology and physics of the oceans.

Hinze, J.O.; *Turbulence,* McGraw Hill, New York, 2nd Edition, 1975, p. 790.
    A comprehensive advanced text.

Ippen, A.I.; *Estuary and Coastline Hydrodynamics,* McGraw Hill, New York, 1966,
    p. 744.  A collection of papers summarizing wave and tide theory,
    harbour resonance and many aspects of estuarine dynamics.

Kinsman, B.; *Wind waves, their generation and propagation on the ocean
    surface,* Prentice-Hall, 1965, p. 676.  A detailed mathematical review
    of wind-wave generation and characteristics.

Lacombe, H.; *Cours d'Océanographie Physique,* Gauthier-Villars, Paris, 1965,
    p. 392.  Covers many of the subjects in the present text (not tides nor
    modelling) but in more mathematical detail.

Lamb, H.; *Hydrodynamics,* Dover (Reprint), 6th Edn., 1932, p. 738.  The basic
    reference on classical fluid dynamics.

LeBlond, P.H. and L.A. Mysak; *Waves in the Ocean,* Elsevier, Amsterdam, in
    press 1978.  A comprehensive, up-to-date, advanced text.

Macmillan, D.H.; *Tides,* American Elsevier, New York, 1966, p. 240.  A non-
    mathematical description of ocean tides, observing instruments, bores,
    etc., with photographs.

McLellan, Hugh J.; *Elements of Physical Oceanography,* Pergamon, 1965, p. 150.
    An introduction to descriptive and dynamic physical oceanography.

Neumann, G. and W.J. Pierson; *Principles of Physical Oceanography,* Prentice-
    Hall, 1966, p. 545.  Moderately advanced text on dynamic and descriptive
    oceanography.

*Oceanography* - Readings from Scientific American; Freeman, 1971, p. 417.  A
    collection of stimulating articles from Scientific American on many
    aspects of oceanography.

Officer, C.B.; *Physical oceanography of estuaries and associated coastal
    waters,* Wiley, 1976, p. 465.  A moderately advanced account of the
    physical theory and applications to typical areas around the world.

Phillips, O.M.; *The Dynamics of the Upper Ocean,* Cambridge University Press,
    1966, p. 261.  A graduate level text on surface and internal waves and
    on oceanic turbulence.

Pickard, G.L.; *Descriptive Physical Oceanography*, Pergamon, 2nd Edn., 1975,
    p. 214. An introduction to descriptive (synoptic) oceanography for
    science undergraduates or graduates.

Pierson, W.J., G. Neumann and R.W. James; *Practical Methods for Observing
    and Forecasting Ocean Waves*, U.S. Naval Oceanographic Office, Publ. 603,
    1955, p. 284. A technical treatise with applications and a chapter on
    wave refraction plotting.

Proudman, J., *Dynamical Oceanography*, Methuen, 1953, p. 409. A basic mathe-
    matical treatise with much detail about waves and tides.

Reid, R.O. (Ed.); *Numerical Models of Ocean Circulation*, National Academy of
    Sciences, Washington, D.C., 1975, p. 364. Proceedings of a symposium
    on numerical modelling. Advanced level but does include several review
    papers on the character of ocean circulation and some discussion on
    directions for future study.

Robinson, A.R. (Ed.); *Wind-driven ocean circulation*, Blaisdell, 1963, p. 161.
    A collection of reprints of papers on the theory of this subject.

Roll, H.U.; *Physics of the Marine Atmosphere*, Academic Press, New York, 1965,
    p. 426. An advanced description of the influence of the sea on the
    atmosphere above it, characteristics and turbulent flow of the atmos-
    phere, thermodynamics.

Russell, R.C.M. and D.M. MacMillan; *Waves and Tides*, Hutchinson, 1952, p. 348.
    Mainly descriptive, with illustrations.

Stern, M.E.; *Ocean Circulation Physics*, Academic Press, 1975, p. 246.
    Geophysical fluid dynamics for graduate physical oceanographers - mathe-
    matical.

Stommel, H.; *The Gulf Stream*, University of California Press, 2nd Edn., 1964,
    p. 248. Both a description of this ocean feature and a review of
    theoretical studies of it. An excellent introduction to physical oceano-
    graphy for upper year undergraduates and graduate students in physics.

Sverdrup, H.U., M.W. Johnson and R.H. Fleming; *The Oceans, their Physics,
    Chemistry and General Biology*, Prentice-Hall, New York, 1946, p. 1087.
    A comprehensive reference text on all aspects of oceanography.

Tennekes, H. and J.L. Lumley; *A first course in turbulence*, MIT Press, 1972,
    p. 300. A first course for graduate students.

Tricker, R.A.R.; *Bores, Breakers, Waves and Wakes*, Elsevier, 1965, p. 250.
    An interesting account of waves near the shore and of bores in rivers.

Turner, J.S.; *Buoyancy effects in fluids*, Cambridge University Press, 1973,
    p. 367. Discusses various consequences of gravity acting on small
    density differences in fluids, e.g., internal waves, instability of
    shear flow, buoyant convection, double diffusion and mixing. Graduate
    level.

Veronis, G.; 'Large-scale ocean circulations' in *Advances in Applied Mechanics, 13*, 1 - 92, 1973. A review of analytic theory with careful attention to approximations used, some discussion of attempts at laboratory simulation of oceanic flows.

Von Arx, W.S.; *An Introduction to Physical Oceanography*, Addison-Wesley, 1962, p. 422. A stimulating introduction to physical oceanography, with special emphasis on current measurements and the use of physical scale models.

Wiegel, R.L.; *Oceanographical Engineering*, Prentice-Hall, 1964, p. 532. Some of the applications of physical oceanographic knowledge, particularly with relation to coastal structures and with emphasis on waves.

## JOURNALS

Recent papers in dynamical physical oceanography may be found in these (and in many others as the reader may find on examining the references):

*Deep-Sea Research*. Pergamon Press, Oxford (since 1953).

*Journal of Geophysical Research* - Oceans and Atmosphere. Amer. Geophys. Union, Washington, D.C. (since 1959). (formerly *Trans. A.G.U.*, see below.)

*Journal of Marine Research*. Sears Foundation for Marine Research, New Haven, Connecticut (since 1939).

*Journal of Physical Oceanography*. Amer. Meteorological Society, Lancaster, Pa. (since 1971).

*Oceanus*. Woods Hole Oceanographic Institution, Woods Hole, Mass. (from 1952). Has short, up-to-date, non-mathematical accounts of recent developments in most aspects of oceanography.

*Tellus*. Stockholm, Sweden (since 1949).

*Transactions*. Amer. Geophys. Union, Washington, D.C. (1920 to 1958).

Two annual reviews of various aspects of oceanography are:

Barnes, Harold (Ed.); *Oceanography and Marine Biology*. An Annual Review. George Allen and Unwin, London. (from Volume 1, 1963).

Warren, B.A. (Ed.); *Progress in Oceanography*, Pergamon Press, Oxford. (from Volume 1, 1964).

TABLES

Sets of numerical data useful in the numerical practice of dynamic oceanography may be found in:

Fleming, R.H.; *Tables for Sigma-t, Journal of Marine Research*, 2, 9-11, (1939). Tables of values of temperature and salinity for whole number of values of sigma-t.

*Handbook of Oceanographic Tables*; U.S. Naval Oceanographic Office, Special Publication 68, Washington, D.C., p. 712, (1966). A collection of tables of use to oceanographers.

*Instruction Manual for Obtaining Oceanographic Data*; U.S. Naval Oceanographic Office Publication 607, Washington, D.C., p. 210, Third Edition (1968), Reprint (1970). A description of routine oceanographic procedures and of standard instruments.

*International Oceanographic Tables*; Unesco, Paris, & National Institute of Oceanography, Wormley, England, Vol. I, p. 128, (1966). For the conversion of conductivity ratio to salinity.

Knudsen, M.; *Hydrographical Tables*, G.E.C. Gad, Copenhagen, p. 63, (1901). Tables for the calculation of sigma-t from values of salinity and temperature.

Lafond, E.C.; *Processing Oceanographic Data*, U.S. Naval Oceanographic Office Publication 614, Washington, D.C., p. 114, (1951). A compilation of tables needed for correcting thermometers, calculating density, specific volume, etc.

*Tables for Sea Water Density*; U.S. Naval Oceanographic Office Publication 615, Washington, D.C., p. 265, (1952). Tables for calculating sigma-t from values of salinity and temperature.

# Index

Acceleration,
   advective,44,83,148,152,179,192,193.
   centripetal,36-38,56,199.
   Coriolis,56,68.
   due to gravity,36,37.
   gravitational,36,37.
   local,216.
Adiabatic change,6,7,23-28.
Advection,140,141.
Advective term,138,142.
Analytic solutions,32,33,91,118,145,
   147.
Angular momentum,12,13,32,122.
Anticyclonic,57.
Axes,
   coordinate,3,32,38,55,214,215.
   fixed,35.
   rotating,35.


Baroclinic,
   compensation,93,126.
   distribution,75,76,109,118,153,158,
      161,163,164,166,167,191,224.
Barotropic distribution,75,76,87,153,
     158,161,193,224.
$\beta$(beta),38,193,194.
   plane,38,111,150,191,193.
Biharmonic operator,114.
Biological production,92,126,141.
Boundary,
   conditions,32,94,104,117,128,137,
     138,141,148,149,161,179.
   eastern,130,131.
   layer,32,90,127-130,133,141.
   western,118,129-131,134,136,151,
     152,154,155,167.
Boussinesq approximation,53,72,86,115.
Brunt-Väisälä frequency,28,53,190.
Buoyancy,24,52,53,72.


Circulation,
   atmospheric,145.
   deep,134-136,140,155,158,165.
   surface,81-83,123,158,160.
   thermohaline,118,134-144,152,166.

wind-driven theory,81-133,166,181.
   Ekman, 81,86-97,101,127.
   Munk,82,114-118,131.
   Stommel,82,110-112.
   Sverdrup,82,97-110.
   world ocean,161-164,181.
Climate, atmospheric,212.
Climatalogical average,114,134,144,150,
   156.
Closure,44,143,149.
Compressibility,23,24,224.
Conservation,
   energy,12.
   heat,12,136-140,142,148,152.
   mass,12,15,16,105.
   salt,136-140,148,152.
   volume,2,15-22,121.
   vorticity,121,136.
Continuity,equation of (see Equation)
*Contra solem,*57
Convergence,21,22,93,105,120,122,141,
   190,196,202,220,221.
Coordinate systems,
   Cartesian,38.
   rectangular,38.
   spherical,38,150,161.
Coriolis
   force,13,14,36,56,57,59,60,68,70,78,
     82,87,89,93,96,97,110,125,126,137,
     153,165,170,191,192,193,224.
   parameter,38,111,166,193,210.
   terms,32,35-37,43,48,50,51,54,80,83,
     86,98,120,127,144,165,191.
C,T,D,6.
*Cum sole,*57.
Current,
   Antarctic Circumpolar,80,161-163,213.
   baroclinic,107.
   barotropic,99,152.
   boundary,155,161.
   Ekman,91.
   equatorial,82,102,117,156.
   geostrophic,77-79,92-96,102,103,134,
     151.
   longshore,175.
   measurement,61.
   North Equatorial,102.